OPTICS
AND
ASTRONOMY

DE DIVERSIS ARTIBUS

COLLECTION DE TRAVAUX
DE L'ACADÉMIE INTERNATIONALE
D'HISTOIRE DES SCIENCES

COLLECTION OF STUDIES
FROM THE INTERNATIONAL ACADEMY
OF THE HISTORY OF SCIENCE

DIRECTION
EDITORS

EMMANUEL
POULLE

ROBERT
HALLEUX

TOME 55 (N.S. 18)

BREPOLS

PROCEEDINGS OF THE XXth INTERNATIONAL CONGRESS
OF HISTORY OF SCIENCE (Liège, 20-26 July 1997)

VOLUME XII

OPTICS

AND

ASTRONOMY

Edited by

Gérard SIMON and Suzanne DÉBARBAT

BREPOLS

The XXth International Congress of History of Science was organized by the Belgian National Committee for Logic, History and Philosophy of Science with the support of :

TABLE OF CONTENTS

Part one
STUDIES IN ANCIENT, MEDIEVAL AND PREMODERN OPTICS

Part two
ASTRONOMY

PART ONE

STUDIES IN ANCIENT, MEDIEVAL AND PREMODERN OPTICS

INTRODUCTION

Gérard SIMON

On trouvera ici les communications qui ont été consacrées à l'optique au cours du XXe Congrès international d'histoire des sciences. Neuf d'entre elles ont été présentées dans un symposium sur " L'optique antique, médiévale et prémoderne (avant Newton) " placé sous la responsabilité d'A. Mark Smith et Gérard Simon. La dixième est une conférence d'ouverture présentée par William Shea, qui porte sur l'interprétation perspective par Galilée des ombres portées des montagnes de la Lune, observées au télescope. Elle s'inscrit tellement dans notre propos, le raisonnement optique de Galilée y est si bien mis en lumière, que nous l'avons intégrée avec joie à nos travaux, heureux de cet apport remarquable et imprévu.

Ces textes présentent une unité thématique qui était celle du symposium. Après une étude sur les fondements de la théorie de la vision chez Ptolémée (A. Mark Smith), on y trouvera deux recherches philologiques sur l'Euclide médiéval, l'une sur la réception arabe de l'*Optique* (E. Kheirandish), l'autre sur la filiation des manuscrits du *De Speculis*, traduction en latin de la *Catoptrique* (Ken'ishi Takahashi). Le reste porte sur la période moderne. Une communication de P. Barker évoque les discussions qui eurent lieu dans les années 1580 entre le Landgrave de Hesse Guillaume IV, Rothmann et Brahé sur la réfraction atmosphérique et sa mesure. On entre déjà dans une période de rénovation et de tension, qu'évoquent M. de Mey et E. de Nil en retraçant la figure d'un grand médecin traditionaliste de la fin du siècle comme Fabricius d'Aquapendente, tandis que S. Unguru et R.D. Chen Morris montrent au contraire en quoi Kepler dans sa théorie de la réflexion se séparait au même moment de la tradition perspectiviste et du conservatisme théorique. Viennent ensuite deux études très complémentaires sur Galilée (W. Shea et P. Hamou), et deux autres sur le rapport d'une part de Descartes à Kepler (G. Simon), d'autre part de Hobbes et des prénewtoniens anglais de Descartes (A. Malet). Tout ceci tisse une trame assez serrée sur la période, ses oppositions, ses lieux critiques et leurs enjeux.

Ce volume aurait dû , hélas, être plus riche encore. Parmi les collègues dont la participation était prévue figurait Wilbur R. Knorr, dont les idées si décapantes ont permis de reposer bien des problèmes que l'on croyait réglés, y compris sur la tradition manuscrite de l'optique antique. Nous comptions beaucoup sur lui pour animer et susciter des débats passionnés, en particulier à propos du corpus euclidien sur lequel il m'avait écrit vouloir intervenir. Malheureusement, la maladie a été plus forte que ses projets et que nos espoirs. Que ce recueil soit aussi un peu comme un dernier hommage rendu au collègue et à l'ami qu'il était devenu.

Il me reste à remercier tous ceux qui ont permis la tenue et la réussite de nos travaux. Outre les intervenants eux-mêmes, le professeur A.I. Sabra, avec qui dans un premier temps ont été discutés le thème et les limites chronologiques du symposium, et avec qui a été esquissée une première liste d'invités. A Mark Smith ensuite, qui a accepté d'en prendre avec moi la responsabilité scientifique. Les organisateurs du Congrès bien sûr, Robert Halleux et son équipe, qui nous ont permis de débattre à loisir, et sur lesquels porte encore la lourde charge de la publication des actes. Et enfin Mme Thill, secrétaire du Centre de recherche " Savoirs et textes " (UMR 8519, CNRS, Universités Lille 3 et Lille 1), qui a assuré le suivi administratif du symposium dans toute sa phase préparatoire.

THE METHODOLOGICAL FOUNDATIONS OF PTOLEMAIC VISUAL THEORY[1]

A. Mark SMITH

Several years ago I suggested that Greek mathematical optics and astronomy shared essentially the same methodological foundations in that both epitomized the endeavor to " save the phenomena "[2]. Accordingly, I argued, both sciences were firmly grounded in the supposition that observed irregularities or anomalies, such as the retrograde motion of planets or the appearance of images behind plane mirrors, were mere illusions. To save the phenomena was therefore to uncover the true regularity behind such illusory appearances. That regularity, in turn, was understood as a function of a saving principle or *salvans* : perfectly uniform circular motion in the case of astronomy, perfectly rectilinear visual radiation in the case of optics. On this basis, classical mathematical astronomy and optics — both brought to perfection by Ptolemy — took shape according to a structure of nesting anomalies and corresponding salvations.

In his 1908 study of the theoretical implications of pre-Galilean astronomy, Pierre Duhem contended that " saving the appearances ", as exemplified in the Ptolemaic model of epicycles and deferents, was an essentially positivist undertaking[3].

1. This paper is an abbreviated version of an article entitled " The Physiological and Psychological Grounds of Ptolemy's Visual Theory : Some Methodological Considerations ", which is currently under consideration by the *Journal of the History of the Behavioral Sciences*.

2. See esp. " Saving the Appearances of the Appearances : The Foundations of Classical Geometrical Optics ", *Archive for History of Exact Sciences,* 24 (1981), 73-100 ; see also " Ptolemy's Search for a Law of Refraction : A Case-Study in the Classical Methodology of 'Saving the Appearances' and its Limitations ", *Archive for History of Exact Sciences,* 26 (1982), 221-240.

3. This seminal study, *ΣΩZEIN TA ΦAINOMENA : Essai sur la notion de théorie physique de Platon à Galilée,* has since been translated into English by Edmund Doland and Chaninah Maschler, *To Save the Phenomena : An Essay on the Idea of Physical Theory from Plato to Galileo,* Chicago, University of Chicago Press, 1969 ; for Duhem's discussion of the instrumentalism of ancient astronomy, see esp. p. 5-24. Bernard Goldstein's discovery of the *Planetary Hypotheses,* in which Ptolemy " physicalizes " the mathematical scheme outlined in the *Almagest,* casts doubt on Duhem's claim ; see Goldstein, *The Arabic Version of Ptolemy's Planetary Hypotheses,* Transactions of the American Philosophical Society 57.4 (1967).

Ptolemy, in other words, never intended his mathematical model to represent cosmological reality. It was meant simply to provide a convenient computational scheme for predicting planetary positions. But what about Ptolemy's attempt to save the optical appearances ? Was that essentially positivist as well ? My response is " no " : Duhem to the contrary, Ptolemy's *Optics* shows a clear commitment to a realist, not a positivist or instrumentalist, position. A brief look at Ptolemy's theory of sight and its physiological and psychological underpinnings will bear this point out.

As Ptolemy describes it, the visual process starts and ends with the *virtus regitiva*, or Governing Faculty. " The origin of nervous activity "[4], in Ptolemy's words, this faculty is the wellspring of sense-perception, serving both to inaugurate the process of sensation and to complete it through perceptual judgment. The Governing Faculty also provides a self-referential platform that lets us make spatial sense of external reality, enabling us not only to distinguish the " outwardness " of visible objects, but to grasp their rightwardness, or leftwardness, or downwardness, or upwardness with respect to our own bodily disposition[5].

In the case of visual perception, the Governing Faculty operates through the two hollow optic nerves that originate at the forefront of the brain and snake through the eye-sockets to the rear of the ocular globes. The mediating entity in all of this is visual flux. Generated in the brain, visual flux passes into and through the two optic nerves, until it reaches the optic chiasma, where the two nerves cross each other. Having crossed and resplit, they channel the visual flux to the eyes, where it enters each ocular globe via the optic nerve and flows to the eye's centerpoint. From here the visual flux streams through the pupil and the corneal surface to form a cone whose vertex defines the center of sight and whose base defines the visual field[6].

If each eye produces its own visual cone and, thus, its own image, then why do we see single rather than double most of the time ? Because, Ptolemy explains, the two eyes normally work in concert to ensure that their respective visual axes intersect at the same point on whatever object is in view. Sharing the same base, the two eyes therefore share the same visual field. Since both images are absolutely congruent, they can be perfectly fused, thus yielding a single representation of the object viewed. What keeps the eyes tracking together is the Governing Faculty, which works through a specific controlling-

4. *Optics* II, 13, in A. Mark Smith, *Ptolemy's Theory of Visual Perception : An English Translation of the* Optics *with Introduction and Commentary*, Transactions of the American Philosophical Society 86.2 (1996), 75.

5. See *Optics* II, 26 and 75, in Smith, *Ptolemy's Theory*, 81-82 and 103.

6. This account of optical anatomy and physiology is based mostly on hints scattered throughout the *Optics*. Ptolemy is explicit about both the sphericity of the eye and the centrality of the visual flux's emanation-point within it. As to the rest of the model, it is suggested by Ptolemy in a variety of ways. For further discussion, see Smith, *Ptolemy's Theory*, 23-30.

point, the Apex, " which lies between [the visual axes] and is where the vertices of the visual cones ought to intersect "[7]. Standing at the vertex of the entire visual system, the Apex is the origin-point of a common axis that cuts the visual axes precisely where they converge on the object during proper binocular vision (see figure 1). It is through the Apex, in short, that the two images are ultimately fused, and its role as the point of image-fusion leaves little doubt that the Apex was meant to lie at the optic chiasma, where the optic nerves unite behind and midway between the eyes.

Altogether, then, the visual flux functions in two interrelated ways : first making physical contact with external objects and, second, forming a perceptual link through which those objects can actually be seen. Now, according to Ptolemy, the proper object of sight is color, and color alone, so when the flux makes physical contact with a visible object, it suffers the " passion " of coloring. More a change of state than a feeling, this passion is transmitted back through the flux to the eye, where it makes an impression on the cornea[8]. Since the rays along which that impression is made are all perpendicular to the corneal surface, the impression itself corresponds point-by-point with the object it represents. Continuing into and through the eye, this impression finally reaches the brain, where the Governing Faculty makes perceptual sense of it.

The visual cone also supplies geometrical clues to the Governing Faculty, and on that basis it is able to make spatial judgments. Distance-perception is a function of the perceiver's sense of ray-length. Size-perception, meanwhile, is based on three things : judgment of the size of the visual angle formed at the center of sight, distance-judgment, and judgment of the object's slant[9]. These are not the sole determinants of spatial perception, however. Color and color-contrasts are crucial. As a result, relative brightness can affect the perception of distance so that, " of objects that lie in the same region, those that are brighter seem closer, ...whereas dim objects appear to be remoter, even if they are actually nearer ". That, Ptolemy goes on to explain, is why " mural-painters use weak and tenuous colors to render things that they want to represent as distant "[10].

Clearly, then, Ptolemy realized that spatial perception cannot be accounted for by ray-analysis alone. Many, if not most, spatial judgments — and misjudgments — that the Governing Faculty makes are in fact based upon non-geometrical information. Much of what we " see " is thus the result of percep-

7. *Optics* III, 35, in Smith, *Ptolemy's Theory*, 144. Ptolemy provides a synoptic explanation of proper binocular vision in *Optics* III, 61, in Smith, *Ptolemy's Theory*, 153.

8. For Ptolemy's discussion of this visual " passion ", see *Optics* II, 23, in Smith, *Ptolemy's Theory*, 79-80 ; for the corneal impression, see *Optics* III, 16, in Smith, *Ptolemy's Theory*, 137-138.

9. On distance-perception through assessment of ray-length, see *Optics* II, 26, in Smith, *Ptolemy's Theory*, 81-82 ; for size-perception, see *Optics* II, 52-63, in Smith, *Ptolemy's Theory*, 92-98.

10. *Optics* II, 124, in Smith, *Ptolemy's Theory*, 120.

tual and intellectual inference rather than of direct visual intuition through geometry. Indeed, as his typology of visual illusions suggests, Ptolemy conceived of seeing as a three-stage inferential process that starts with brute sensation, and progresses through perception to intellectual apprehension. This tripartite perceptual scheme is reflected in a hierarchy of visual illusions. At the lowest level, the visual flux suffers passions, such as " breaking " (i.e. reflection or refraction), that cause a misapprehension of the object's location. At the second level, that of perception, such things as color-perspective can cause misapprehension of shape or distance. And, finally, at the highest level, where intellectual judgments are made, false conclusions about size can arise from a variety of miscues. The most salient example of such misjudgment is the so-called Moon Illusion[11].

The importance of psychological intervention in his account of sight is manifest in Ptolemy's analysis of image-formation in spherical concave mirrors[12]. Let ABC in figure 2 represent the mirror. Let D be its center of curvature, E the center of sight, EB the incident ray, O (or O_1) the point-object to be seen in the mirror, and BO the reflected ray. Therefore, O's (or O_1's) image, I (or I_1), will be located at the intersection of EB and DO. However, when the object lies at point H, there is no intersection between the extension of incident ray EB and cathetus of reflection HD. Yet, despite the geometrical imperatives, we must see something in this case, because the reflected flux actually reaches the object along BH. Accordingly, Ptolemy assures us, " the image of [H] will seem to coalesce [with the surface] where point B lies on the mirror, and it will take on both its position and color. For there is no proper, determinate location for the image of anything located at... H "[13]. The point is this : since it must somehow see an image of H, even though the geometrically-derived location is indeterminate, the visual faculty projects an image to a falsely determinate location — the mirror's surface. Why the need for this ghost image ? Because the nebulous confusion that we actually see on the mirror's surface under the prescribed circumstance cannot be adequately explained by ray-geometry ; psychological agency must be appealed to. For Ptolemy, therefore, the demands of observed fact overrode the dictates of ray-analysis.

What, then, was the ontological status of the Ptolemaic visual ray ? Here the contrast between Ptolemy and Euclid is profoundly instructive. For Euclid, on the one hand, the visual ray was absolutely real — a reified line-of-sight[14]. For

---

11. For Ptolemy's overall analysis of visual illusions, see *Optics* II, 83-142, in Smith, *Ptolemy's Theory*, 106-128. On the so-called Moon Illusion, see *Optics* III, 59, in Smith, *Ptolemy's Theory*, 151.

12. The analysis that follows derives from Ptolemy's account in *Optics* IV, 71-73, in Smith, *Ptolemy's Theory*, 194-195.

13. *Optics* IV, 72, in Smith, *Ptolemy's Theory*, 195.

14. For Euclid's construction of the visual cone and his consequent definition of visual rays, see the first three postulates of Euclid's *Optics* in *Euclidis Optica,... cum scholiis antiquis* in Heiberg and Menge (eds), *Euclidis opera omnia*, vol. 7, Leipzig, Teubner, 1895, 2.

Ptolemy, on the other, it was a mere analytic convention, a virtual rather than a real entity. It would seem therefore that Euclid was the realist and Ptolemy the instrumentalist. But in fact the opposite is true. Perhaps the clearest evidence of this point is to be found in Euclid's account of how visual acuity decreases with distance. His rationale is simple : the farther an object gets from the vertex of the visual cone, the fewer rays strike it until, finally, it disappears into the gap between rays[15]. But such reasoning, Ptolemy rejoins, flies in the face of experience. For one thing, if there really were inter-radial gaps, then we would perceive the world as a sort of mosaic rather than as a continuum. Worse, it follows from Euclid's account that even the remotest objects would reappear and disappear by turns as rays passed over them during a visual sweep. Therefore, concludes Ptolemy, since our apprehension of external reality is continuous, the visual cone must be continuous. The ray, in other words, is a mathematical fiction, not a physically real entity[16].

By denying the ray physical reality, Ptolemy was in no way abandoning realism in favor of positivism. On the contrary, his conception of the ray was modified to fit rather than mold experience. Indeed, his realism is evident in the pains he took throughout the *Optics* to accommodate his theory, as well as his analytic structure, to actual observation — hence his continual recourse to empirical examples as well as to experimental confirmation. Accordingly, Ptolemy regarded ray-geometry as a supremely useful analytic tool, but one with distinct limitations — as witness his treatment of the ray as a virtual entity and his refusal to bow fully to the dictates of ray-analysis in explaining image-formation in concave mirrors. Furthermore, Ptolemy broadened not only the scope, but also the theoretical basis of visual analysis to include a full range of physiological and psychological considerations. Ptolemy, in short, meant his account of visual perception to be as coherent and as comprehensive as possible — and thus as reflective as possible of the physical reality it was intended to explain.

The contrast I have drawn between Ptolemy, the realist, and Euclid, the positivist, has significant implications for our understanding of ancient optics in general. The canonical view harks back to David Lindberg's account in *Theories of Vision from Alkindi to Kepler*[17]. According to him, ancient optics developed along three essentially separate lines. The first, which is philosophical, can be traced from the presocratics to the later Stoics and Epicureans. The second, which is medical, found its endpoint in Galen. The third, which is mathematical, passed from Euclid to Ptolemy, who " not only extended Euclid's mathematical analysis of vision, but also enlarged it to include additional phys-

---

15. For Euclid's reasoning here, see propositions 2 and 3 of Euclid's *Optics* in Heiberg, *Euclidis Optica*, 4-6.

16. See *Optics* II, 50-51, in Smith, *Ptolemy's Theory*, 91-92.

17. Chicago, University of Chicago, 1976.

ical, physiological, and psychological elements "[18]. Ptolemy's work, in short, was essentially an offshoot of Euclid's.

But, in fact, Ptolemaic optics represented more than a mere extension of previous work. For a start, although both Euclid and Ptolemy relied upon ray-analysis, they conceived of the ray in fundamentally different ways : Euclid saw it as *the* analytic determinant ; Ptolemy as *one* analytic determinant. Ptolemy, moreover, did not just add physical, physiological, and psychological elements to a Euclidean nucleus. Ptolemy's *Optics* represents a seamless integration rather than a mere aggregation of those elements. Lindberg's three traditions thus come to full convergence in the *Optics*. This is significant because it suggests that, as seems to be the case with the *Almagest*, Ptolemy's *Optics* marks a significant departure from, not just a continuation of, tradition[19].

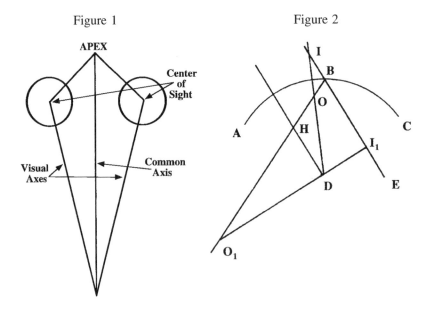

Figure 1   Figure 2

18. *Theories of Vision*, 52.

19. For recent revisionary views of Ptolemaic astronomy and its originality, see A.C.Bowen, " The Art of the Commander and the Emergence of Predictive Astronomy ", forthcoming in N. Fox and Ch.J. Tuplin (eds), *Science in Ancient Greece*, Oxford, New York, Oxford University Press and B.R. Goldstein, " Saving the Phenomena ", *Journal for the History of Astronomy*, 28 (1997), 1-12.

WHAT " EUCLID SAID " TO HIS ARABIC READERS : THE CASE OF THE *OPTICS*

Elaheh KHEIRANDISH

Euclidean Optics has a complex textual history which has recently become the subject of thorough examination and discussion. Indeed, the textual tradition of Euclidean optics is so problematic that a simple designation such as " Euclid's Optics " cannot even be used meaningfully as a single point of reference. For a long time, such a designation referred to a text published in various early European editions (Zamberti, 1505, Pena, 1557, David Gregory, 1703, to name but a few). At about a century ago, it came to refer to *Heiberg's Euclid*, a different text based on the manuscripts of a Greek *Optika* that J.L. Heiberg edited as *Euclidis Optica*, as distinguished from the previously known set of Greek manuscripts this time presented in the same edition with the title *Opticorum recensio Theonis*, as a recension by the fourth-century commentator, Theon of Alexandria (Heiberg, 1895) [as represented by a 12th century Vienna codex and 10th century Vatican codex respectively][1]. More recently, the re-examinations of the Greek Euclidean tradition by Alexander Jones and Wilbur Knorr (in what has left less and less of Euclid in the former and more and more of it in the latter), has left us with *Euclid's Optics* according to *their* respective standpoints[2].

The case of the Arabic tradition of Euclidean optics is no less problematic. It is not simply the case that there *was* such a text as " Euclid's Optics " from the standpoint of the Arabic tradition. Neither are the textual problems limited

---

1. J. Heiberg, *Euclidis Optica, Opticorum Recensio Theonis, Catoptrica, cum scholiis antiquis in Euclidis Opera Omnia*, Heiberg and Menge (Eds), 1895, vol. VII : *Optica*, 1-121 ; *Recensio Theonis*, 143-247 ; French translation, P. Ver Eecke, *Euclide, l'Optique et la Catoptrique*, Œuvres traduites pour la première fois du grec en français, avec une Introduction et des Notes par Paul Ver Eecke, Nouveau Tirage, Paris, Albert Blanchard, 1959 ; English translation, H.E. Burton, " The Optics of Euclid ", *Journal of the Optical Society of America*, 35 (1945), 357-372.

2. A. Jones, " Peripatetic and Euclidean Theories of the Visual Ray ", *Physis*, 31, n° 1 (1994), 47-76 ; W.R. Knorr, " Pseudo-Euclidean Reflections in Ancient Optics : A Re-Examination of Textual Issues Pertaining to the Euclidean *Optica* and *Catoptrica* ", *Physis*, 31, n° 1 (1994), 1-45.

to there being more than one Arabic version bearing the name of Euclid and the title of *Optics* (*Kitāb Uqlīdis fī Iktilāf al-manāzir* and *Kitāb al-Manāzir li-Uqlīdis*), or these not having an exact or consistent correspondence to the two main Greek versions as expected[3]. In the case of the Arabic tradition, there is also another text passing under the name of Euclid (*Kitāb al-Mir'āh li-Uqlīdis*, better known as the Latin *De speculis*)[4], with yet another formulation of the first three definitions of the *Optics*, and one that the problematic transmission of the latter's opening lines in both Greek and Arabic, seems to have given a particularly crucial role.

The title of the present paper, by itself, captures some sense of the problems involved through each of its individual components : " What Euclid *said* to his Arabic readers " is a problem which, at least in " The Case of the *Optics* " and the standing issues of identification and attribution, must be addressed in terms of what was said in the name of Euclid, in this case from the standpoint of the Arabic tradition. A major problem is also posed by *what* was said to the Arabic readers, whether by Euclid or in his name, since this, as the title already reveals, was not exactly what was said to the Greek readers. Yet another problem has to do with *Euclid*, this time as a reference not to the author, but to the exact source of his supposed statements : " Euclid said " (*qāla Uqlīdis*), appearing next to more than one formulation of the opening definitions of the *Optics* in Arabic, refers not just to the Euclid of the *Kitāb al-Manāzir*, where the Arabic version of the Euclidean definitions literally translate into the Arabic " version " of Euclidean visual theory[5] ; the reference seems to be also directed to the Euclid of *Kitāb al-Mir'āh*, in whose name some early Arabic texts present *their* version of the visual-ray hypotheses with an almost word-for-word correspondence with what we now know as the [pseudo-] Euclidean *De speculis*.

The Arabic tradition has much to contribute to all three problems, formu-

3. E. Kheirandish, *The Arabic Version of Euclid's Optics : Kitāb Uqlīdis fī Ikhtilāf al-manāzir*, Edited and Translated with Historical Introduction and Commentary, 2 vols, New York, Springer Verlag, *Sources in the History of Mathematics and Physcal Sciences* (forthcoming).

4. *Tractatus [pseudo-] Euclidis De speculis*, A. Björnbo and S. Vogl, *Alkindi, Tideus und Pseudo-Euklid : Drei optische Werke, Abhandlungen zur Geschichte der mathematischen Wissenschaften*, 26, 3 (1912), 1-176, 97-173. On the Arabic version, see A. I. Sabra, " A Note on Codex Biblioteca Medicea Laurenziana Or. 152 ", *Journal for the History of Arabic Science*, 1 (1977), 276-284, where it is distinguished from the pseudo-Euclidean Catoptrics extant in Greek ; see K. Takahashi, *The Medieval Latin Traditions of Euclid's Catoptrica*, A Critical Edition of *De speculis* with an Introduction, English Translation and Commentary, Higashi-ku, Kyushu University Press, 1992 ; on the problematic transmission of the opening Euclidean definitions with reference to this text, see E. Kheirandish, *The Arabic Version of Euclid's Optics : Kitāb Uqlīdis fī Ikhtilāf al-manāzir*.

5. E. Kheirandish, " The Arabic 'Version' of Euclidean Optics : Transformations as Linguistic Problems in Transmission ", *Tradition, Transmission, Transformation : Proceedings of Two Conferences on Pre-modern Science, Held at the University of Oklahoma*, ed. F.J. Ragep and S.P. Ragep with S. Livesey, Leiden, Brill, 1996 (Collection de travaux de l'Académie internationale d'histoire des sciences, t. 37).

lated here as the " who " (author), the " what " (statement), and the " which " (source) problems, for the highly problematic case of Euclidean optics. This is a case where the pressing problems of attribution and transmission have already formed the subject of recent studies on both its Greek and Arabic traditions, and the problems of identification as discussed here for the first time, are to point to the directions for further study.

The evidence from the Arabic Euclidean tradition is of immediate interest to the standing problems of attribution. Not only is there such a text as " The *Optics* of Euclid " from the standpoint of the Arabic tradition, the name of Euclid being explicitly and repeatedly attached to the Arabic versions passing under the title of *Optics ;* this is a text which, even as the product of a transmission involving at least two versions, is passed on as the single voice of Euclid. There is no direct association with the name of Theon of Alexandria with reference to Arabic optics, and no proof that the Greek Euclidean version long supposed to be a " recension " by Theon, ever existed in Arabic, despite clear traces of it in both the Arabic and Latin traditions[6] ; the only other *Optics* being Ptolemy's, this time known to have existed in Arabic at some point, though with an apparently limited circulation[7].

What the Arabic tradition has to offer to the complex problems of transmission, is of particular interest, and again, much beyond that of the Arabic tradition itself : on the one hand, what was said to the Arabic readers of the *Optics* was not exactly what was said to the Greek readers in either of the Greek versions now associated with Euclid's name, and this all affected what was, in turn, said to the Latin readers. On the other hand, what passed on as Euclid's words in the most critical part of the text, was not only representing the formulation of the Euclidean visual theory in the *Optics* (*Kitāb al-Manaẓir/ Ikhtilāf al-manāẓir li-Uqlīdis*) already transformed through a chain of flexible terms and interchangeable scripts ; the [pseudo-] Euclidean *De Speculis* (*Kitāb al-Mir'āh li-Uqlīdis)* is another source from which subsequent " versions " seem to have been drawn.

For the purposes of the present paper, the alternative formulations of the hypotheses of Euclidean visual theory are presented in two Appendices :

---

6. W.R. Theisen, *The Mediaeval Tradition of Euclid's Optics*, Ph.D. dissertation, University of Wisconsin, 1972, facsimile, University Microfilms, 1984 [unpublished], 324, n. 10, suggests one of the four Latin versions : *Euclidis de aspectuum diversitate* to be a translation from such an Arabic version.

7. A. Lejeune, *L'Optique de Claude Ptolémée dans la version latine d'après l'arabe de l'émir Eugène de Sicile,* Édition critique et exégétique augmentée d'une traduction française et de compléments, Leiden, E. J. Brill, 1989 (Collection de travaux de l'Académie internationale d'histoire des sciences, t. 31) ; A. M. Smith, " Ptolemy's Theory of Visual Perception : An English Translation of the *Optics* with Introduction and Commentary ", *Transactions of the American Philosophical Society Held at Philadelphia for Promoting Useful Knowledge*, 86, 2 (1996), 1-270.

On the limited circulation of Ptolemy's *Optics* in Islamic lands, see Sabra, *The Optics of Ibn al-Haytham*, vol. 2, LXIV-LXXIII.

Appendix I contains the wording of the Greek, Arabic, and Latin versions of the opening Euclidean definition, their variants and modern translations, and Appendix II, the formulation of the same in a variety of Arabic works, both early and late, with various degrees of correspondence to the problematic opening lines of the *Optics*. Appendix II also contains full bibliographical references for the manuscripts or editions, and an abbreviated list of previous treatments of each text in chronological order, since this is not always fully documented.

The complex transmission of the *Optics* is best demonstrated by the notable case of the very first Euclidean definition [Appendix I]. Not only is this the most critical part of the Euclidean text in terms of presenting *the* statement of the visual-ray hypothesis ; it is indeed the most problematic part in terms of textual transmission. It is in the same part of the text that the two main Greek versions are most divergent. It is also in the same part, though not in the same manner, that the variance of two slightly different Arabic versions, and in fact all Arabic variants, is most striking. Again, it is the same part that has been considered defective in the Greek, and the same part, for which restoration efforts can be documented in Arabic. It is, no surprise, the same part where subsequent translations and interpretations are also most problematic.

From the wordings of the Greek and Arabic versions of the opening definition, it can be observed that the opening Arabic term corresponding to the most central term for the Euclidean visual theory, namely *opseis* [in def. 1 of R and def. 2 and 3 of both O and R] meaning visual rays, becomes one which is not only singular, but no longer exclusive to visual radiation [*al-shu'ā'* in all Arabic variants]. Perhaps more critical, the description of the Greek visual rays as having a great magnitude [*megethôn megalôn* only in O] becomes, in its Arabic translation, a description which can at once refer to an infinitely great magnitude *or* multitude (*lā nihāyata li kathratihā*). The description itself is applied no longer to the visual rays, but to the produced rectilinear paths or directions (*sumūt*) [apparently corresponding to *diastêma*, itself used in different senses in O and R], such that these paths or directions [or intervals as in R] are infinitely great, no longer in magnitude, but in multitude.

In the complex case of the opening definition, it is even enough to pay close attention to its closing portion, a portion passed on to all the manuscript copies of the early and late Arabic versions of that definition. The expression *megethôn megalôn*, employed in the Greek original [only in O] to describe the manner of visual radiation, is, by itself, problematic enough to have translated into modern languages in various forms : as *des grandes grandeurs* by Ver Eecke, " of great extent " by Burton, " indefinitely " by Cohen and Drabkin, and " great magnitude ", by both Knorr and Jones. Great magnitude is also what this can be read as, in its Arabic translation. But the subtle point is that it is not all that the expression " *lā nihāyata li kathratihā* " can mean in Arabic. In fact, not only is great multitude a much more common reading, we know that it is as such that the expression has been subsequently passed on : the ara-bic expression reappears as " *multitudini non est finis* " in the latin version of Euclid's *Optics* entitled *Liber de aspectibus*, and as *multitudine infinitae* in the extant Latin text of al-Kindī's *De aspectibus*, where even in the absence of the original, it is clear than an infinite number is applied to " straight lines extended from the rays' point of origin "[8]. Ibn al-Haytham continues to speak of " light issuing from the object to the surface of the eye on an infinity (*bi lā nihāya*) of variously situated straight lines " and Kepler, of " lines [of light] infinite in number issuing in every point in the visual field "[9]. In this way, the slightest shift in the transmission of even a single term, that from magnitude to multitude, was enough to have produced the multiplying effect, the compo-nent so critical to the transition from the single cone of the Euclidean visual model to the multiple-point radiation models of some later figures.

In the next definition, that of the visual cone [not included in Appendix I], the origination point of the ray, whether as a single ray or a cone of rays [now in multitude, according to one sense of the Arabic version] is not in the eye [ἐν τῷ ὄμματι/en tôi ommati in O], but " next to " or " on " the eye [*yalā al-'ayn/* or *'alā al-'ayn* depending on the MSS., which is close to πρὸς τῷ ὄμματι/*pros tôi ommati* (near/at the eye) in R [a direct translation from here would have been *fī al-'ayn* (inside the eye), with another close orthography in Arabic]. Together, the combined elements transform the original Euclidean assumptions of straight lines traveling an interval [i. e. distance] of great magnitude [in O], or else visual rays traveling along straight lines, while making some interval from one another [in R] to a model of an infinite multitude of paths/directions produced by the conic collections of rays, now issuing from the eye's outer surface[10].

There are other components here and elsewhere to have contributed to what seems like an almost instantaneous transformation of Euclidean optics by vir-

---

8. *Liber de aspectibus*, ed. W. R. Theisen, *The Mediaeval Tradition of Euclid's Optics*, 336 : *Radius egreditur ab oculo super lineas equales rectas et accidit post ipsum rectitudo recta cuius multitudini non est finis* ; *De aspectibus*, prop. 11f, see A. Björnbo and S. Vogl, *Alkindi, Tideus und Pseudo-Euklid : Drei optische Werke*, 15 : *Si enim a puncto sinapis lineae rectae protrahe-rentur, egrederentur ab eo lineae in multitudine infinitae*.

9. Ibn al-Haytham, *Risāla fī al-Ḍaw* ", *Majmū' al-rasā'il, Hyderabad*, 1357 (=1938-39), 10 : " light issues from the object to the surface of the eye on an infinity of (bi lā nihāya) variously situated straight lines ", translation of passage in Sabra, *The Optics of Ibn al-Haytham, Books I-III On Direct Vision*, Translated with Introduction and Commentary by A. I. Sabra, 2 vols, London, The Warburg Institute, University of London, 1989, 10 (Studies of the Warburg Institute, vol. 40, I-II). See also Rashed, " Le 'Discours de la Lumière' d'Ibn al-Haytham (Alhazen) ", *Revue d'His-toire des Sciences*, 1968, 21 :197-224. Kepler, *Ad Vittellionem Paralipomena*, p. 157, reference given in D. C. Lindberg, *Theories of Vision from Al-Kindi to Kepler*, Chicago, London, The Uni-versity of Chicago Press, 1976, 193, n. 75.

10. This model is geometrically comparable to the principle of punctiform analysis of light radiation as described in Lindberg, *Theories of Vision from Al-Kindi to Kepler : Alkindis Critique of Euclid's Theory of Vision* ; 30, 226, and Sabra, *The Optics of Ibn al-Haytham*, vol. 2, 23 ; see also Kheirandish, " The Arabic 'Version' of Euclidean Optics : Transformations as Linguistic Problems in Transmission ".

tue of its Greco-Arabic transmission alone. And there are signs of textual defects in the case of both the Greek and Arabic traditions, which are likely to have affected the successive transmission of the texts most central concepts. The evidence to that effect includes a missing opening page from one of the Arabic variants of the *Optics* (dated 600H) as well as an explicit statement at the colophon of another variant (dated 692H) that its original was " poor, disturbed, and cut-off " (*wa kāna aṣluhu saqīman, muqlaqan, bitran*), a statement, which though not specific enough in terms of the exact place or source of the problem, supports the assumption of textual defects, strongly suggested by the particularly problematic opening lines in both Greek and Arabic[11].

Finally, what was transmitted of the founding assumptions of Euclidean visual theory to the Arabic readers, was beyond the *Optics* itself, as there was also what was said of those very same assumption to the Arabic readers of a *De speculis*, also bearing Euclid's name in its Arabic version. The statements on Appendix II, present at least two distinct formulations of the hypotheses of Euclidean visual-theory, one involving a ray (*al-shu'ā'*) issued (*yukhraj*) from the eye (*al-'ayn*) producing paths (*sumūt*) of infinite (*lā nihāya*) multitude (*kathra*), and in the form of a cone *makhrūṭ*, with apex (*ra's*) on the eye ('*alā al-'ayn*), and base at the object (*mubṣar*) (I), and the other, involving a luminous power (*quwwa nūriyya*) spreading (*yanbathth*) from the eye's pupil (*nāẓir*) in a pinelike (*ṣanawbariyya*) form, its pointed corner (*zujj= mustaḥadd*) being next to the pupil (*nāẓir*) and its base widening (*ittasa'at*) as it gets further away (*ba'uda*) (II).

From the statements on Appendix II, it is immediately clear that most of the treatments surviving from the 3rd/9th to the 7th/13th centuries (those numbered V-VIII) have a close correspondence to the formulation of the Arabic *Optics* (I), and in fact to the same variant of it (the first in Appendix I) : note the repeated occurrence of the terms *al-shu'ā'* (ray), *yukhraj* (is issued), *'ayn* (eye), *sumūt* (paths), *lā nihāya* (infinite), *kathra* (multitude), *makhrūṭ* (cone), *ra's* (apex), and *mubṣar* (visible object) ; while at least two treatments, both from the earlier centuries (III-IV), are unmistakably comparable to the formulation in the Arabic *De speculis* (II), the most striking cases now being, *ṣanawbar* (pine), *zujj, mustaḥadd* (corner) and *nāẓir* (pupil). Indeed, so close is the correspondence of the non-standard terminology of Euclidean optics in the latter case[12],

---

11. On the evidence from the Arabic tradition, see E. Kheirandish, *The Arabic Version of Euclid's Optics : Kitāb Uqlīdis fī Ikhtilāf al-manāẓir*, vol. 1, XXVI-XXVII, XXIX, and vol. 2, 6 ; on the case of the Greek tradition, see A. Jones, " Peripatetic and Euclidean Theories of the Visual Ray ", 52, and Knorr , " Pseudo-Euclidean Reflections in Ancient Optics ", 29, n. 48-49.

12. Note that the term *zujj*, used as a non-standard form for the cone's apex in the Arabic *De speculis* and Aḥmad ibn 'Īsā's Manāẓir wa al-marāya, may have been also intended for a slightly different transcription in the unique manuscript of the otherwise identical corresponding text of al-Kindī's *Taqwīm*. The term appears as *raḥb* in J. Jolivet and R. Rashed, *Œuvres Philosophiques et Scientifiques d'Al-Kindī*, vol. 1 : *L'Optique et la Catoptrique* in *Islamic Philosophy Theology and Science*, Leiden, Brill, 1997, 163 (Texts and Studies, vol. XXIX).

that it is as if Euclid's *Kitāb al-Mir'āh* (*De speculis*) acted as a supplement to Euclid *Kitāb al-Manāẓir* (*Optics*) for the critical case of the opening lines we now know as having involved textual defects and restoration efforts in both its Greek and Arabic traditions. There is certainly more to be said about this important historical problem, a task demanding a proper study of the early Arabic texts to which I hope to contribute with my forthcoming publication of the full text of Aḥmad ibn 'Īsā's *Optics and Burning Mirrors* " in the tradition of Euclid ". This is only one of a few cases revealing that the problem with what " Euclid said " (*qāla Uqlīdis*) to the Arabic readers of his *Optics*, is not just whether this *was* in fact what was said by Euclid himself, and why it is not exactly *what* was said to his Greek readers, but also, that it is not always reducible to what Euclid said in his *Optics*.

APPENDIX I : GREEK, ARABIC AND LATIN VERSIONS OF THE OPENING
EUCLIDEAN DEFINITION OF THE *OPTICS*

I. GREEK VERSIONS :

Euclidis Optica [= O, Heiberg, (1895), p. 2] :
Ὑποκείσθω τὰς ἀπὸ τοῦ ὄμματος ἐξαγομένας εὐθείας γραμμὰς φέρεσθαι διάστημα μεγεθῶν μεγάλων.
[Burton (1945), p. 357 : " Let it be assumed that lines drawn directly from the eye pass through a space of great extent " ; Ver Eecke (1959), p. 1 : *Supposons que les lignes droites qui émanent de l'œil se propagent à divergence des grandes grandeurs* ; Cohen and Drabkin (1958), p. 257 : " Let it be assumed that the rectilinear rays proceeding from the eye diverge indefinitely " ; Jones (1994), p. 52 : " Let it be postulated that the straight lines drawn out from the eye travel an interval of great magnitude " ; Knorr (1994), p. 29 : " Let it be postulated that the straight lines leading out from the eye are carried out a distance of great magnitude "].

Opticorum Recensio Theonis [= R, Heiberg (1895), p. 154] :
Ὑποκείσθω τὰς ἀπὸ τοῦ ὄμματος ὄψεις κατ' εὐθείας γραμμὰς φέρεσθαι διάστημα τι ποιούσας ἀπ' ἀλλήλων.
[Ver Eecke (1959), p. 57 : *Supposons que les rayons visuels émanés de l'œil se propagent suivant des lignes droites faisant quelque divergence entre elles* ; Jones (1994), p. 52 : " Let it be postulated that the visual rays from the eye travel along straight lines, while making some interval from one another " ; Knorr (1994), p. 29, n. 48 ; " Let it be postulated that the visual rays from the eye are carried in straight lines that make a certain distance from each other "].

II. Arabic versions :

Kitāb Uqlīdis fī Ikhtilāf al-manāẓir [Kheirandish (1998), vol. I, p. 2-3] : " A ray (al-shu'ā') [or a collection of rays] is issued from the eye in straight lines, while producing straight paths of infinite multitude ".

الشعاع يخرج من العين على خطوط مستقيمه ومعه يحدث سموت مستقيمة لا نهاية لكثرتها

Variant : [Visual rays=] ... issued...in unequal and unequally distant straight lines...

[الناظر من العين هى أنّب الشعاع يخرج من العين على خطوط مستقيمة

مختلفة الابعاد والمقادير وتحدث به سموت مستقيمة لا نهاية لكثرتها

III. Latin versions :

Euclid, *Liber de aspectibus* [Theisen (1972), p. 336] : *Radius egreditur ab oculo super lineas equales rectas et accidit post ipsum rectitudo recta cuius mutitudini non est finis.*

Appendix II : alternative arabic formulations
of the opening euclidean definition of the *Optics*

I : The Book of Euclid on Optics (*Kitāb Uqlīdis fī Ikhtilāf al-manāẓir*).
Istanbul : Topkapi Sarayi, Ahmet III 3464, 4 : 59b, 625H.
[Krause (1936), p. 441 ; Sezgin (1974), p. 117 ; Kheirandish (1991, 1996, 1998) ; Rashed (1997), p. 10].

الشعاع يخرج من العين على خطوط مستقيمه ومعه يحدث سموت مستقيمة لا نهاية لكثرتها

II : The Book of Catoptrics by Euclid = [Pseudo-] Euclidean *De speculis* (*Kitāb al-Mir'āh li-Uqlīdis*).
Florence : Medicea Laurenziana Or. 152, 2, 1266 (unique ?, incomplete), fol. 104a
[Sabra (1979), p. 283, Kheirandish (1991), p. 132, Kheirandish (1998) ; Rashed (1997), p. 337 ; Björnbo and Vogl (1912) : p. 97-173 ; Theisen (1972), p. 276].

العين ينبثّ من ناظرها قوّة نوريّة تؤثر فيما لاقت من الجوّ أجمع ضياء صنوبريّاً

زُجّه أعنى مستحدّه عند الناظر نفسه وكلّما بَعُد اتّسعت قاعدته

III : The Book on Optics and Burning Mirrors Composed by Aḥmad ibn 'Īsā in the tradition of Euclid on Demonstrations for Vision (*Kitāb fīhi al-Manāẓir wa al-marāyā al-muḥriqa ta'līf Aḥmad ibn 'Īsā 'alā madhhab Uqlīdis fī 'ilal al-baṣar*).
Istanbūl : Süleymaniye, Ragip Paşa 934, undated (ca. 600 H ?), fol. 5b-6a.

[Krause (1936), p. 513-514 ; Kheirandish (1991), p. 18-22 and Appendix ; Kheirandish (1996, 1998) ; Sabra (1989), v. 2, p. XXXII, n. 31 ; p. XXXVI-XXX-VII, n. 39 ; Rashed (1997), section on Burning Mirrors, Appendix III, p. 649].

قالت الفلاسفة وأقليدس منهم ومعهم انَّ

العين ينبثَ من ناظرها قوَّة نوريَّة تؤثر فيما لاقت من الجوَّ المضيء أعنى الهواء

إذا كان مضيناً أجمع ضياء شكله صنوبرياً شبيه بالزُّجَّ

ورُجَهَ أعنى مستحدهَ عند الناظر نفسه وكلَّما بَعُد اتَّسعت قاعدته

IV : The Book of Abī Yūsuf Ya'qūb ibn Isḥāq al-Kindī to Some of his Friends on the Errors and Problems of Euclid in his Book Called al-Manāẓir (*Kitāb Abī Yūsuf Ya'qūb ibn Isḥāq al-Kindī ilā ba'ḍ ikhwānihi Fī Taqwīm al-khaṭā' wa al-mushkilāt allatī li-Uqlīdis fī Kitābihi al-mawsūm bi al-Nāẓir [al-Manāẓir]*).

Qum : Ayatollah Mar'ashī-yi Najafī, 7580, 69b-102b, 960H (unique ?), fol. 69b.

[Mar'ashī (v. 19), p. 390 ; Rashed (1977), p. 162-335].

العين ينبثَ من ناظرها قوَّة نوريَّة تؤثر فيما لاقت من الجوَّ

أجمع ضياء صنوبرياً رحباً؟ ثرُزُجَهَ؟ب أعنى مستحدهَ عند الناظر نفسه وكلَّما بَعُد اتَّسعت

قاعدته

V : The Book on the Reasons for what Occurs in Mirrors on Account of the Difference of Aspects (Kitāb fī 'Ilal mā ya'riḍu fī al-marāyā min ikhtilāf al-manāḫẓir 'alifahu ... Qusṭā ibn Lūqā al-Yūnānī).

Mashhad : Āstān-i Quds 392 (5593), 1 : 1-31, 867H (unique ?), p. 5 ; [Gulchīn Ma'ānī (1350=1971), v. 8, p. 344 ; Toomer (1976) p. 26-27 ; Kheirandish (1991), p. 23-24 and Appendix ; Kheirandish (1996, 1998) ; Rashed (1997), p. 572-646].

البحر يكون بجحاع ينبثَ من الجحن ويجع عج المبجرات فتُبجَر بالجحاع الواقع علجها

,,, والجحاعات الواقجة عج المبجرات تكون كثجرة عج جهات مختلجة ,,,

VI : al-Ya'qūbī, Ta'rīkh (1883), v. 1, p. 139.

[Krause (1974), p. 441 ; Kheirandish (1996, 1998) ; Rashed (1997), p. 42].

ولأقليدس هذا كتاب فى المناظر واختلافها من مخارج العيون والشعاع يقول فيه

إنَّ الشعاع تخرج من العين على خطوط مستقيمة وتحدث بعد سموت لا نهاية لكثرتها

VII : Taḥrīr Kitāb al-Manāẓir li-Uqlīdis

Tehran : Muṭahharī (Sipahsālār) 4727, 7 : 97b-101b (186-194), 671H, fol. 97b

[Dānishpazhūh and Munzavī (11340=1962), v. 3, p. 349 ; Kheirandish (1998)].

الجحن تُحدث ,,, شجاعاً كما تُحدثه الأجرام النجَّرة وحدها بجحنه

ويكون ذلج الجحاع كجنَّه منبجثٌ من الجحن وخارجٌ منها ثمَّ إنَّه يجحر آلة لها فى الإبجار

فتختلف أحوال المناظر لاختلاف أوضاعه فلجّجدّق بجلجَ ولجُتوهّم ذلج الجّجاع

متّجلاً بالجّجن عج خطوطٍ مستجّجمة ولّجُحدث سموتاً مستجّجمةً لا نهايةَ لكثرتها

VIII : Tajrīd Uqlīdis fī al-Manāẓir
by Jamāl al-Dīn Muḥammad ibn al-Ṣaḥib Kamāl al-Dīn ʿUmar ibn Aḥmad
ibn Hibatallāh (Ibn Abī Jarāda)
Cairo : Dār al-Kutub, DR 638 : 12 fols., ca. 1100H (unique ?), 1b.
[King (1979) p. 453 ; King (1986a), v. 2, p. 1030 ; King (1986b), p. 24 ;
Kheirandish (1991, 1998) ; Rashed (1997), p. 10].

الشعاع يخرج من العين على خطوط مستقيمه ويحدث سموتاً لا نهاية لكثرتها

Abbreviated titles in the Appendices :

Björnbo and Vogl (1912) = *Alkindi, Tideus und Pseudo-Euklid : Drei
optische Werke,* herausgegeben und erklärt von A. Björnbo und Sebastian
Vogl, *Abhandlungen zur Geschichte der mathematischen Wissenschaften,*
Leipzig, Berlin, 26, 3 (1912), 1-176.

Burton (1945) = H.E. Burton, " The *Optics* of Euclid ", *Journal of the Optical Society of America,* 35 (1945), 357-372.

Cohen and Drabkin (1958) = M. R. Cohen and I.E. Drabkin, *A Source Book
in Greek Science,* Cambridge, Harvard University Press, 1958.

Gulchīn Maʿānī (1350=1971) = *A Catalogue of Manuscripts in the Astan
Quds Razavi Library (Fihrist-i Kutub-i khaṭṭī-yi Kiṭabkhānih-i Āstān-i Quds-i
Raḍavī)* by Ahmad Golchin Maani, v. 8, Mashhad, A publication of the Cul-
tural and Library Affairs, n° 6, 1350=1971.

Heiberg (1895) = J.L. Heiberg, *Euclidis Optica, Opticorum recensio Theonis, Catoptrica, cum scholiis antiquis,* edidit Heiberg, *Euclidis Opera Omnia,*
ediderunt J.L. Heiberg et H. Menge, vol. VII, Leipzig, Teubner, 1895, 1-121.

Heiberg (1895) = J.L. Heiberg, *Euclidis Optica, Opticorum recensio Theonis, Catoptrica, cum scholiis antiquis,* edidit Heiberg, *Euclidis Opera Omnia,*
ediderunt J.L. Heiberg et H. Menge, vol. VII, Leipzig, Teubner, 1895, 143-247.

Dānishpazhūh and Munzavī (1340=1962) = *Catalogue of Manuscripts in
Sepahsālār-Library* by M.T. Daneshpazhuh and A.N. Monzavi *(Fihrist-i
Kitābkhānih-i Sipahsālār),* 3 vols (v. 3-5), Tehran, Iranian Council for Philos-
ophy and Humanistic Studies, v. 3, ed. Munzāvī, 1340=1962.

Jones (1994) = A. Jones, " Peripatetic and Euclidean Theories of the Visual
Ray ", *Physis,* 31, 1(1994), 47-76.

Kheirandish (1991) = E. Kheirandish, *The Medieval Arabic Tradition of
Euclid's Optika,* Ph.D. Thesis [unpublished], Harvard University, 1991.

Kheirandish (1996) = E. Kheirandish, " The Arabic 'Version' of Euclidean
Optics : Transformations as Linguistic Problems in Transmission ", *Tradition,
Transmission, Transformation : Proceedings of Two Conferences on Pre-mod-*

ern Science, Held at the University of Oklahoma, ed. F. Jamil Ragep and Sally P. Ragep with Steven Livesey, Leiden, E. J. Brill, 1996, 227-247.

Kheirandish (1998) = E. Kheirandish, The Arabic Version of Euclid's *Optics* : *Kitāb Uqlīdis fī Ikhtilāf al-manāẓir*, Edited and Translated with Historical Introduction and Commentary, 2 vols, New York, Springer Verlag, forthcoming (Sources in the History of Mathematics and Physical Sciences).

King (1979) = D.A. King, " Notes on the Sources for the History of Islamic Mathematics ", *Journal of the American Oriental Society*, 1979, 99.3, 450-459.

King (1986a) = D.A. King, *A Catalogue of the Scientific Manuscripts in the Egyptian National Library*, Cairo, General Egyptian Book Organization in collaboration with The American Research Center in Egypt, and the Smithsonan Institution, Part II, 1986 [Part I, 1981].

King (1986b) = D.A. King, *A Survey of the Scientific Manuscripts in the Egyptian National Library*, Published for The American Research Center in Egypt, Catalogs, v. 5, Winona Lake, Indiana, Eisenbrauns, 1986.

Knorr (1994) = W.R. Knorr, " Pseudo-Euclidean Reflections in Ancient Optics : A Re-Examination of Textual Issues Pertaining to the Euclidean *Optica* and *Catoptrica* ", *Physis*, 31, 1(1994), 1-45.

Krause (1936) = M. Krause, " Stambuler Handschriften islamischer Mathematiker ", *Quellen und Studien zur Geschichte der Mathematik, Astronomie und Physik*, Berlin, 3, 4 (1936), 437-532.

Mar'ashī (v. 19) = S.A. Ḥusaynī, , and S.M. Mar'ashī, *Fihrist-i Nuskhih'hāyi khaṭṭī-yi Kitābkhānih-i Ḥaḍrat-i Āyatullāh al-'Uẓmā Najāfī-yi Mar'ashī ...* , v. 11, 1364 (=1985) ; v. 12, 1365 (=1986).

Rashed (1997) = R. Rashed, J. Jolivet, *Œuvres Philosophiques et Scientifique d' Al-Kindī*, vol. I : *L'Optique et la Catoptrique* in *Islamic Philosophy Theology and Science*, Leiden, E. J. Brill, 1997 (Texts and Studies, vol. XXIX).

Sabra (1989) = A.I. Sabra, *The Optics of Ibn al-Haytham* : *Books I-III On Direct Vision, Translated with Introduction and Commentary by A.I. Sabra*, 2 vols, London, The Warburg Institute, University of London, 1989 (Studies of the Warburg Institute, vol. 40 I-II).

Theisen (1972) = W.R. Theisen, *The Mediaeval Tradition of Euclid's Optics*, Ph.D. Thesis, University of Wisconsin, 1972, facsimile, Ann Arbor, University Microfilms International, 1984.

Toomer (1972) = G.J. Toomer, *Diocles on Burning Mirrors* : *The Arabic Translation of the Lost Greek Original*, Edited with English Translation and Commentary by G.J. Toomer, New York, Springer-Verlag, *Sources in the History of Mathematics and Physical Sciences*, n° 1, 1976.

Ver Eecke (1959) = P. Ver Eecke, *Euclide, l'Optique et la Catoptrique*, Œuvres traduites pour la première fois du grec en français, avec une Introduc-

tion et des Notes par Paul Ver Eecke, Nouveau Tirage, Paris, Librairie Scientifique et Technique Albert Blanchard, 1959.

EUCLID'S *DE SPECULIS*

ITS TEXTUAL TRADITION RECONSIDERED[1]

Ken'ichi TAKAHASHI

De speculis is a Latin translation of Euclid's *Catoptrica* which was translated into Latin probably in Sicily in the mid-twelfth century. The large number of extant manuscripts of *De speculis,* of which at least 52 MSS are known, indicates its wide circulation in the Middle Ages. To take into account the circulation of other works on optics will be helpful for understanding the popularity of *De speculis*[2]. *De speculis* is only second in the number of copies produced to Pecham's *Perspectiva communis,* which was the most popular standard textbook in the Middle Ages. Moreover, our *De speculis* is one of the earliest optical works translated into Latin.

I published in 1992 a book which contains three different versions of Euclid's *De speculis*[3]. During my stay at University of Wisconsin in 1996-97, I have had an opportunity to examine seven additional manuscripts which were new to me and to reconsider *De speculis'* textual tradition. So I would like to report certain aspects of my research on a book on catoptrics, focusing mainly on the two aspects of *De speculis'* textual tradition : first on the modification that a certain text undergoes when translated from one language into another ; and second on the issues concerning the anonymous translator of *De speculis.*

1. I would like to express my gratitude to Prof. Lindberg, University of Wisconsin-Madison, USA, for providing me with a number of microfilm copies of *De speculis.*

2. The numbers of extant manuscripts of the several major works on optics are as follows : Euclid's *De speculis* (52 MSS), Euclid's *De visu* (34 MSS), pseudo-Euclid's *De speculis* (18 MSS), Hero of Alexandria's *De speculis* (2 MSS), Tideus' *De speculis* (15 MSS) ; Ptolemy's *Optica* (15 MSS) ; Alhazen's *De aspectibus* (20 MSS) ; Grosseteste's *De iride* (12 MSS) ; Roger Bacon's *Perspectiva* (39 MSS), Roger Bacon's *De multiplicatione specierum* (26 MSS) ; Witelo's *Perspectiva* (25 MSS), and Pecham's *Perspectiva communis* (64 MSS). This information is based on Lindberg's *A Catalogue of Medieval and Renaissance Optical Manuscripts*, Toronto, 1975.

3. K. Takahashi, *The Medieval Latin Traditions of Euclid's Catoptrica,* Kyushu University Press, 1992.

At the outset, let us look at an enlarged version of manuscripts consulted in my study which shows a stemma of manuscripts used. In Fig. 1, I show the origin of each manuscript in its main outline. The letters are the abbreviations of the manuscripts used. The seven manuscripts, which I have newly consulted, are indicated by the encircled abbreviation with two letters. I do not enter the problem of how to establish these relationships. It should be noted, however, that my stemma is intended only to identify the main source of influence for each manuscript, and second that there are probably several generations between a manuscript and its closest derivative in my stemma.

Fig. 1 : Stemma of Manuscripts Used

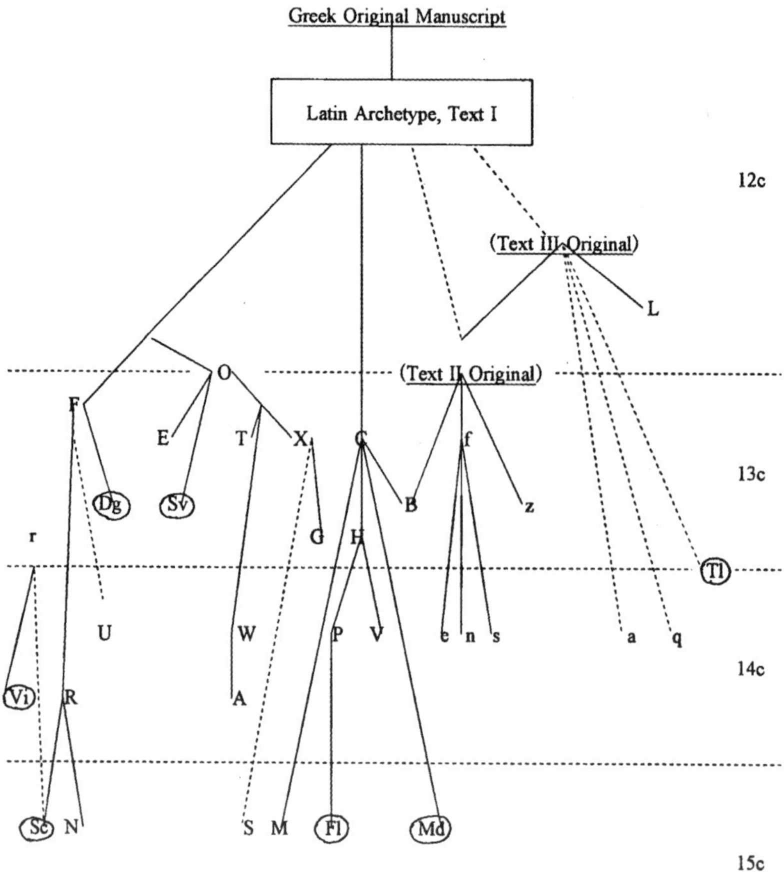

Manuscripts newly consulted :

Dg = Oxford, Bodleian Library, MS Digby 174, fols. 179r-181v. (13th c.)

Fl = Foligno, Biblio. Jacobilli, MS 54 (A. II. 27), fols. 2v-1 Iv. (15th c.)

Md = Madrid, Biblio. Nac., MS 9119, fols. 307r-310v. (15th c.)

Tl = Toledo, Archivo y Bibl. Capitulares de la Cathedral, MS 98.22, fols. 87v-89v. 13-14th c.)

Sc = Schagl, Stiftsbibl., MS 126 (824.236), fols. 130v-150v. (1466)

Sv = Seville, Bibl. Colombina, MS 7.6.2, fols. 54 (55) v-63 (64) v. (13th c.)

Vi = Vienna, Osterr. Nat. Bibl., MS 5210, fols. 88r-95r. (14th c.)

In order to understand the first aspect, that is, the translation and interpretation process, I would like to examine in the following order : first, the original Latin translation which is Text I ; and second, the paraphrased Latin versions which are Texts II and III and others.

First let me make some short comments on the Greek original itself. Scholars have doubted the authenticity of the *Catoptrica*. However, recent work by Knorr, Jones, and myself verifies its authenticity[4]. The work seemed to have a lot of false assertions and deficiencies and lacked the mathematical rigor, unworthy of the author of the masterpiece of Greek mathematics, entitled the *Elements*. To mention just one example among many for the sake of my argument below, Prop. 7 seemed to assert or presume the same contents as Prop. 16 which states the location of images in plane mirrors. In other words, the preceding proposition seemed to be dependent on the subsequent proposition. This is an example that allegedly lacks mathematical rigor. However, closer scrutiny of the theoretical contents of these as well as others, shows that the so-called false assertions, deficiencies, and lack of mathematical rigor resulted from their misunderstanding. The rehabilitation of Euclidean authenticity itself is an interesting story, but the time does not permit me to go into this topic here. In any case, it is worth noting that the Greek original seemed to be far from Euclidean perfection as exemplified in his *Elements*. Its lack of sophistication encouraged some intelligent medieval readers to expand the text with their own comments.

THE LATIN TRANSLATION FROM THE GREEK (TEXT I)

Now let us go to the examination of Text I. I have consulted 37 manuscripts in my study, of which Text I forms the biggest group, comprising 25 MSS. However, all the medieval Latin manuscripts of *De speculis* have distinctive features in common which can be traced back to Text I. First, they have thirty

4. W.R. Knorr, " Archimedes and the Pseudo-Euclidean Catoptrics : Early Stages in the Ancient Geometric Theory of Mirrors ", *Archives Internationales d'Histoire des Sciences,* 35 (1985), 53-73 ; A. Jones, " On Some Borrowed and Misunderstood Problems in Greek Catoptrics ", *Centaurus,* 32 (1987), 1-17.

two propositions in all, whereas the Greek text edited by J.L. Heiberg has only thirty. The discrepancy between the Greek and the Latin numerations of the propositions can easily be explained : Proposition 28 of the Greek text has in the proof four cases, the last two of which become distinct propositions in the Latin text as Propositions 29 and 30. Thus the Latin text has the last two propositions of the Greek text numbered as Propositions 31 and 32. To repeat, regardless of other discrepancies between the texts, the existence of this common feature shows beyond any doubt that Text I was the common source of other texts.

Second, our work was translated into Latin from Greek, not from Arabic. Translations were usually made from the twelfth century on mainly from either Greek or Arabic texts. It should be noted that Text I has some such transliterations of Greek words as *fantasia, theoremati, apodixibus, cathethus* and *periferia*. Moreover, this fact is further supported by a comparison of the Greek and the Latin texts and by lack of any Arabicisms. The translation faithfully follows the word-for-word translation technique which is a prominent feature of twelfth century translations. This fact does not necessarily mean that the twelfth century translators were incompetent in their works. To cite C.H. Haskins' expression, this derives from their belief that " The texts which these scholars rendered were authorities in a sense that the modern world has lost, and their words were net to be trifled with "[5]. The most outstanding example is the use of the *id quod* construction. *Id quod* means " that which ". And this is used when a modifier, other than a simple adjective, stands in the attributive position in the Greek. A following example of this sort is enough to show this Grecism of the translation in our work : " ἐπὶ τῆς δίχα καὶ πρὸς ὀρθὰς τεμνούσης τὴν ἐκ τοῦ κέντρου (*super eam que in duo equalia et perpendiculariter secantem eam que e centro*) ". The translation shows that the Latin text faithfully reflects the order of the Greek words which are sandwiched within an *id quod* construction.

THE PARAPHRASED VERSIONS (TEXTS II AND III, AND SOME MSS)

Now let us go to the paraphrased versions, that is, Texts II and III, and some other MSS. And I would like to bring light to the way in which the text was received in the Middle Ages.

It is generally true that, in the Middle Ages, several Latin versions were often found, which were adapted from a particular Latin text. In the medieval Latin tradition, Euclid's *Catoptrics* shared almost the same fate as that of his *Elements*. The *Elements* has three medieval Latin versions, (1) the translation, (2) the abridged version, and (3) the commentary, which are respectively

---

5. C.H. Haskins, *Studies in the History of Mediaeval Science*, Harvard University Press, 1924, 151-152.

named Adelard I, Adelard II and Adelard III by Prof Clagett for Adelard who allegedly translated it. Our texts II and III differ from Text I just as Adelard II or III of Euclid's *Elements* does from Adelard I. The difference is found only in the proofs proper, not in the postulates and the enunciations of the propositions except for minor variants and omissions. MS *Tl* is a rare exception in that it sometimes rewrites enunciations of the propositions into more natural Latin as well as the proofs proper.

Among the two Latin texts, let us look only at Text III. This text is preserved in a single MS *L*, which is incomplete, having the enunciations and proofs of propositions 1-6 and, after that, the enunciation of proposition 16. This indicates the most important feature, that is, a rearrangement of the order of propositions. The original of Text III probably had, immediately before proposition 7, propositions 16-18 which teach how to determine the location of images produced by the plane, the spherical concave and spherical convex mirrors. In fact, we have the complete versions of this type in MSS *a, q,* and *Tl*. As I mentioned before, Prop. 7 seemed to presuppose Prop. 16 even until recent times, though this interpretation was nothing but a misunderstanding from the historical point of view. Even though this interpretation was correct in the historical sense, the rearrangement of propositions nonetheless shows the logical acuteness on the part of the scholar who wrote this Text III. In passing, I would like just to mention his mathematical acuteness in his comments on Postulate 3. My presentation at this symposium is deeply stimulated by the early existence of this peculiar paraphrased version III.

Now it is time that I should go to the second aspect of *De speculis'* textual tradition which is concerned with the issues of its anonymous translator. Before going to its examination, it is necessary to do some stylistic analyses of Text III because each scholar has his own style which he reveals in his text, especially in paraphrased versions. By doing so, I hope that stylistic analysis will shed new light on the anonymous translator, and my case study will contribute to the elucidation of the activities of medieval scholars in general and the production of a more nuanced picture of them.

A STYLISTIC ANALYSIS OF TEXT III

Text III has some characteristics in addition to the author's effort to give more natural Latin to the text. We can point out at least three noteworthy features.

The first feature in Text III's style is the regular use of the stock phrase for the equality of angles of incidence and reflection. In its typical form, Text III has *reverberatio fiat ad pares angulos,* whereas the law of reflection is expressed in Prop. 1 of Texts I and II in the following manner : *In planis speculis et convexis et concavis visus in equalibus angulis revertuntur.* In the former expression, emphasis should be put on two words, *reverberatio* and *pares,* which are rarely found in medieval optical treatises. To my knowledge, the first word is found only in Ptolemy's *Optica* which was translated by Eugene the Emir in the middle of the twelfth century, while the second appears in *De radiis visualibus* which is a translation of Euclid's *Optica* made from Arabic.

The second feature is the following phrase to express proportions : *que est proportio A ad B ea est C ad D.* It is noteworthy that this expression is used in the comments to Postulate 3 although the postulate itself employs such expression as *sicut A ad B ita C ad D.*

The third is the consistent change of *" non iam "* into *" iam non ".* The former is, of course, the translation of " οὐκέτι " in Text I.

ISSUES CONCERNING THE ANONYMOUS TRANSLATOR OF *DE SPECULIS*

Before going directly to the problem of the anonymous translator of *De speculis,* let us first pay attention to the codex that contains Text III[6]. Euclid's two optical works, *De visu* and *De speculis,* which are contained in it, were written either before the first half of the 13th century or even much earlier, which means the second half of the 12th century[7]. Therefore this codex has one of the oldest extant texts of Euclide's two works. Reading the text of *De visu,* we can conclude that the author is the same person as the author of Text III, because the text of this *De visu* has the same characteristic features as Text III we have examined before[8].

The text of this *De visu* has an interesting and significant feature as well : in addition to the Greek enunciation of the propositions, the text has another enunciation for some propositions which are preceded by the phrase : *Habet alia translatio.* In fact they are the enunciations of *De radiis visualibus,* as already pointed out by Prof. Theisen in his doctoral thesis[9].

From these facts we can conclude first that the author of Text III knew very well Ptolemy's *Optica* translated by Eugene the Emir around the middle of the twelfth century, second that he also knew the *De radiis visualibus* whose translator is not identified yet[10], and third that he was not the translator of Euclid's

6. MS Add.17368 (London, British Library). The following is all the contents that the codex has : fols. 1r-14r (Bemard of Verdun, Astronomy) ; fols. 15r-35v (Anonymous, Astronomical Tables) ; fols. 36r-51v (Johannes de Persorio, Astronomical Calendar) ; fols. 52r-59v (John Pecham, *Perspectiva communis*) ; fols. 60r-69r (Euclid, *De visu*) ; fol. 69v (Archimedes, *De quadratura circuli*) ; fols. 70r-71r (Euclid, *De speculis*) ; fols. 71v-74v (blank).

7. This is the judgement of Dr. Coutenay, professor of history at the University of Wisconsin-Madison, who examined a reprint of this codex at my request.

8. To mention a few examples, in Props. 19-22 of *De visu,* one can find the same expressions used for the law of reflection and the proportional relationships as used in Text III. *Iam non* is also used in Prop. 3.

9. W.R. Theisen, *The Mediaeval Tradition of Euclid's Optics,* Ann Arbor, Ph. D Dissertation, 1972, p. 17 fn. 37, p. 58.

10. D.C. Lindberg, *Theories of Vision from Alkindi to Kepler,* The University of Chicago Press, 1976, 211.

Optica and *Catoptrica,* because his texts were not the translations, but the Latin paraphrased versions, including some enunciations of *De radiis visualibus* in the case of *De visu.* Moreover, it is highly probable that he studied at Sicily, because Eugene the Emir lived and worked there.

Now it is time to turn to the anonymous translator's famous preface to the *Almagest* which is available in the reproduction of the preface in Haskins' book[11]. According to the preface, this translator came to Sicily from Salerno in order to get access to Ptolemy's *Almagest* after having heard the news of its copy brought by an emissary, by name Aristippus, as a present from the Greek emperor to the Sicilian king. The crucial passage of the preface is as follows :

...Cum occulte quidem alia, manifeste vero mens scientie siderum expers prefatum mihi transferre opus prohiberent, grecis ego litteris diligentissime preinstructus, primo quidem in Euclidis Dedomenis, Opticis, et Catoptricis, Phisicaque Procli Elementatione prelusi. Dehinc vero prefatum Ptolemei opus aggressus expositorem propicium divina providente Eugenium, virum tam grece quam arabice lingue peritissimum, latine quoque non ignorarum, illud contra viri discoli voluntatem latine dedi orationi...[12]

It is evident from the preface first that mention was made to three Euclidean minor works (the *Data, Optica,* and *Catoptrica*) as well as Proclus's *Elementatio physica,* and second that the translator alluded to two persons, mentioning Eugene the Emir who was, in his opinion, " most skilled in Greek and Arabic and not unfamiliar with Latin ", and the other uncertain " ill-tempered man ". However, his wording is so ambiguous that we are informed of almost nothing about the nature of his study of these four works. The key word is, of course, *prelusi.* Heiberg took it to mean " translated " while Haskins more cautiously understood it to mean that " he not only studied them but tried his hand (*prelusi*) at turning them into Latin "[13]. At any event, they shared the opinion that the anonymous translator of the *Almagest* also translated four minor works mentioned above.

Although there are many studies on the identification of the anonymous Sicilian translator of Ptolemy's *Almagest* (mid-twelfth century) and his scholarly activities, it suffices for us to mention just two extreme opinions : according to Prof. Murdoch, *Almagest* and *Elements* were the works of one translator, while Euclid's other three works (*Data, Optica,* and *Catoptrica*) were translated by another scholar[14]. However, following Heiberg, Haskins and Boese, Prof Ito claims that these five texts were translated by the same scholar[15].

11. C.H. Haskins, *op. cit.,* 191-193.

12. *Ibidem.* Underlines added.

13. J.L. Heiberg, " Noch einmal die mittelalterliche Ptolemaios-Uebersetzungen ", *Hermes,* XLVI (1911), 209 ; C.H. Haskins, *op. cit.,* 179.

14. J.E. Murdoch, " Euclides Graeco-Latinus : A Hitherto Unknown Medieval Latin Translation of the *Elements* Made Directly from the Greek ", *Harvard Studies in Classical Philology,* 71 (1967), 249-302.

15. S. Ito, *The Medieval Latin Translation of the Data of Euclid,* University of Tokyo Press, 1980, 26-38.

Their arguments are based on the analyses of the translation technique of certain Greek words. Let me show you in a tabulated form such Greek and Latin words that are relevant to their arguments and mine (see table 1).

Before examining the table, however, some explanatory remarks are necessary for understanding it. The table gives the Latin words for the Greek word written in the first column. In collecting the information of the Latin words, I consulted Prof. Murdoch's paper for the Greco-Latin version of the *Elements*. As for the four minor works, I benefited respectively from Ito's edition of *Data*, Theisen's edition of *De visu*, my edition of *De speculis* and Boese's edition of *Elementatio physica*[16]. Moreover, I used also Lejeune's text of Ptolemy's *Optica* and Theisen's doctoral thesis which contains the text of *De radiis visualibus* in order to obtain additional information[17]. The number that follows each Latin word shows the frequency of its appearance in each text. Square bracket [] shows the number given by Prof. Murdoch while parenthesis () shows my counting of its appearance. My counting is different from and more accurate than that of Prof. Murdoch.

Now let us examine the table itself. First let us look at the last four entries from " δίχα " to " ὥστε ". My tabulation suggests that all six works in question show almost the same pattern of distribution of the Latin words. In this sense, these works were the products of translators who shared the common scholarly heritage. This fact would lend support to the legitimacy of talking about the Sicilian " school " of translators[18]. However, a closer look at the table would reveal some predilections of individual translators. The table suggests that the first two works were not translated by the same person as the other four. To repeat the main points of Prof. Murdoch's argument, the Latin words for three Greek words, " ὅτι ", " δή " and " ἐπιζεύγνυμι ", are noticeably different between the two groups. As for " ὅτι ", the first two words exclusively use *quoniam* whereas the latter four use *quod* as well as *quoniam,* showing some fluctuation between them. We will come back to this point later. As for " δή " it should be noted that *ergo* becomes dominant in the first two tracts while the word is completely absent in the latter four. And in passing, please pay attention to the Latin words for " ἀλλά ". *Verum* found in the first two is not found in the succeeding three works even though it is used in the last work.

16. S. Ito, *op. cit.* ; W.R. Theisen, " *Liber de visu :* The Greco-Latin Translation of Euclid's *Optics* ", *Medieval Studies,* 41 (1979), 44-105 ; K. Takahashi, *op. cit.* ; Helmut Boese, " Die Mittelalterliche Uebersetzung der ΣΤΟΙΧΕΙΩΣΙΣ ΦΥΣΙΚΗ des Proclus ", Berlin, Akademie-Verlag, 1958.

17. A. Lejeune, *L'optique de Claude Ptolémée dans la version latine d'après l'arabe de l'émir Eugène de Sicile,* E.J. Brill, 2nd ed., 1989 ; W.R. Theisen, *op. cit.,* 403-422.

18. As will be shown later, the Sicilian " school " of translators consists at least of the following four persons : Henricus Aristippus, Eugene the Emir, the translator of *De radiis visualibus,* and the translator of the *Elements* and the *Almagest.* The third person might be one of the senior members of this school who was refered to as " the ill-tempered man " by the last person, the so called Sicilian anonymous translator.

Finally the translation of " ἐπιζεύγνυμι " shows again the decisive difference of translation technique : the exclusive use of *copulo* in the first two works. The existence of these differences seems to me convincing enough to accept Prof. Murdoch's position concerning the issues of the anonymous translator of the *Almagest*. Moreover, in order to buttress this position, I would like to mention the argument of Prof. Busard who published the Greco-Latin version of the *Elements*. He insists that there may have been two translators of the *Elements*, the one who translated Books I-XI and XV, and the other who translated Books XII-XIV[19]. He also states that the former translator was not the person who translated *Data*, which is not easy to reconcile with Prof. Ito's argument.

So, if we follow Prof. Murdoch's line of thought which concludes that the translator of the *Almagest* did not translate the Euclidean minor three works, what then can we claim concerning his involvement with these and other works ? In order to present my tentative conclusion concerning this problem, let us recapitulate the facts that we have mentioned so far, including some plausible ones.

(1) Text I of *De speculis* is a Latin translation made from the Greek original.

(2) Text III of *De speculis* is a paraphrased version of Text I.

(3) The author of Text III knew both Ptolemy's *Optica* translated by Eugene the Emir and the *De radiis visualibus* translated from the Arabic text of Euclid's *Optica*.

(4) In his preface the translator of the *Almagest* mentioned Eugene by name and a certain " ill-tempered man ".

(5) He applauded Eugene for his competence in Greek and Arabic languages. (In passing, I would like to add that Eugene's native language was Greek).

(6) After learning the Greek (*preinstructus*), the translator of the *Almagest prelusi* Euclid's *Data, Optica* and *Catoptrica* and moreover Proclus's *Elementatio physica*.

(7) From stylistic analysis, the *Elements* and the *Almagest* were translated by the same scholar.

(8) This scholar did not translate the Euclidean three minor tracts nor Proclus's *Elementatio* physica.

With these facts in mind, I would like to elaborate Prof. Murdoch's following suggestion : " Here, one is tempted to speculate : if, with Haskins, we closely connect the translator's citation of these four works with his immediately following mention of Eugene the Emir, could it not be that Eugene translated them and that these Latin versions were studied by our anonymous

19. H.L.L. Busard, *The Medieval latin Translation of Euclid's Elements*, Stuttgart, 1987, 2-4.

scholar ? "[20] We now have greater pieces of evidence than the mere mention of Eugene.

It is beyond doubt that the anonymous translator of the *Almagest* was closely related with Eugene who was, as it were, a senior member of the Sicilian " school " of translators. It is generally true that a newcomer begins his study by first imitating and absorbing the achievements of seniors and then elaborating them. As for the interpretation of *prelusi*, I would like to urge the necessity to interpret it in closer connection with *preinstructus*. As Haskins has already pointed out, our anonymous translator of the *Almagest* did not have a good command of Greek at that time, because he said *preinstructus*, not *iam instructus*[21]. Even after learning the Greek, it would have been necessary for him to do some preliminary study of other minor works at least for two reasons : the first reason is linguistic, because getting a better command of Greek, especially obtaining and improving the translation skills, was well beyond the mere acquaintance with Greek language, the second reason is scientific, because knowledge of these minor works was prerequisite for translating the *Almagest* on account of their belonging to the intermediate course between the *Elements* and the *Almagest*. This kind of preliminary study was best pursued by making a paraphrased version of each Latin text translated by someone else, in this case, probably by Eugene. As was the case at that time, this kind of study probably includes such activities as adding the comments to the text, rewriting the original proofs and the like in the way as shown in our Text III. In short, from my point of view, *prelusi* does not necessarily involve the translating activity.

Moreover, since we know that Eugene was skilled in both Greek and Arabic and that he translated Ptolemy's voluminous *Optica* into Latin from Arabic, it would be possible to consider that he also translated from Greek at least some minor works on optics. This would have been much easier for him, because Greek was his mother tongue.

This possibility would be raised to the plausibility by pointing out a close relationship between the Latin words used in the translations of the *Almagest* and the other Euclidean minor works. Let us take as an example the exclusive employment of *quoniam* and *copulo* in the first two works. As we have seen in the case of translation of the Greek " ὅτι ", the latter four works fluctuated as to the choice of a suitable Latin word, *quod* or *quoniam*. It is noteworthy that *quod,* which would be a natural Latin translation for " ὅτι " is more frequently employed than *quoniam* in *Data*, but the latter becomes dominant in the other works, as can be seen in the table. This state of affairs is understandable if we consider the restriction that *quod* for " ὅτι ", would suffer from the *id-quod* construction. Consider an example of the following sentence in Prop. 94 of

20. J.E. Murdoch, *op. cit.*, 298, note 94.
21. C.H. Haskins, *op. cit.*, 160, note 12.

Data : " λέγω... ὅτι τὸ ὑπὸ τῶν ΑΔ, ΕΖ χωρίον δοθέν ἐστιν " (*Dico... quo-niam quod sub AD et EZ spatium datus est*). The appearance of *quod* as a result of *id-quod* construction would prohibit the possible introduction of *quod* as a suitable word for " ὅτι ". We can see the same *id-quod* construction in the *Almagest*. Therefore, the fluctuation between *quod* and *quoniam* seems to me to reflect the difficulty and its tentative solution by the first generation of Sicilian school of translators. As for the exclusive use of *copulo*, we know that this word was frequently employed in Eugene's translation of Ptolemy's *Optica* and rarely used in *De visu*.

In short, we can find a close relationship between the translator of the *Almagest* and Eugene even in their translation technique. We can safely state that the former person benefited greatly from the latter. Therefore, I would like to propose the following as a plausible hypothesis : the translator of the *Almagest* studied Euclidean minor three works in their Latin translations made by Eugene the Emir (our Text I included), probably with Greek texts at his side with the intention of his future translation of the *Almagest* ; and his preliminary study of these works was handed down to us, at least for *De speculis* and *De visu*, in the paraphrased version contained in MS *L* (our Text III). And this hypothesis will also provide us with a clue to a seeming enigma that the paraphrased version of *De speculis*, namely Text III, existed at an earliest phase of the *De speculis'* textual tradition.

Table 1 : Some Translated Latin Words

Greek	Elements	Almagest	Data	De visu	De speculis	Elem. Phys.	Ptolemy's Optica	De radiis visualibus
ὅτι	quoniam [58]	quoniam [30]	quod (98) quoniam (9) quia (1)	quoniam (24) quod (19)	quoniam (10) quod (9)	quoniam (24) quod (4)		
δή	ergo [46] et [4] autem [1] itaque [1]	ergo [20] vero [2] igitur [1] itaque [1]	autem (13) et (10) vero (7) enim (1) at (1)	autem [11] vero [2]	vero (7) autem (1) quoque (1)	autem (6) vero (5)		
ἀλλά	sed [14] verum [9] vero [1]	sed [9] verum [6] at [1]	sed (42) atqui (1)	sed [11] vero [1]	sed (5)	sed (24) verum (4)		
οὐκέτι	non adhuc [1]	?	non iam (3) iam non (1)	non iam (3)	non iam (3)	non iam (4)		
ἐπιζεύγνυμι	copulo [?]	copulo [?]	coniungo (23) produco (1)	coniungo (46) copulo (1)	coniungo (12) traho (7)	\<none\>	protraho (108) copulo (76) produco (75)	duco (13) produco (2) iungo (2)
ἄγω	?	protraho (?) duco (?)	traho (70) protraho (34)	?	traho (6) duco (4)	\<none\>	coniungo (10) duco (8)	

γράφω	scribo [?]	scribo [?]	describo (9)	scribo (2)	describo (1)	\<none\>	describo (13)	figuro (2)
ἀναγράφω etc.	?	?	describo (43)	describo (31)	describo (2) duco (2)	\<none\>	scribo (1)	
δίχα		in duo equa (?) in duo equa-lia (?)	in duo equa (5) in duo (4)	in duo equa (2) in duo equa-lia (2)	in duo equa (4) in duo (1) in duo equa-lia (1)	in duo equa-lia (3)	?	per equalia (1)
πάλιν	rursum [13] rursus [5]	rursum [38] rursus [2]	rursum (15)	rursum (7)	rursum (8) rursus (4) item (1)	rursum (4) rursus (2)		
ὡς... οὕτως	ut... ita [32] sicut...ita [10]	sicut...ita [6] ut...ita [5]	sicut...ita (105) ut... ita (2) quemad-modum... ita (1)	sicut...ita (12)	sicut... ita (5) quemad-modum... ita (1)	sicut...ita (1)		
ὥστε	quare [15] ut [1]	quare [39] ita ut [5] ut [4]	quare (47) ut (7) quasi (2)	quare (10) unde [1]	quare (14) ut (2)	quare (3)		

BRAHE AND ROTHMANN ON ATMOSPHERIC REFRACTION

Peter BARKER

My object in this brief paper will be to make a preliminary case for the re-evaluation of two assumptions about the earliest work on atmospheric refraction. The first of these is the common claim that Tycho Brahe (1546-1601) was responsible for the earliest work on this phenomenon — and especially the earliest refraction table. I will show that another important German observatory was also conducting refraction studies leading to the construction of a correction table that may well predate Brahe's. The second assumption is that atmospheric refraction was interesting primarily as a (small) correction to the coordinates of celestial objects in positional astronomy. I will show that refraction was seen by both Brahe and his contemporaries as direct evidence about one of the most contentious issues of his day : the reality of celestial spheres. Those who rejected celestial spheres consistently appealed to measurements of atmospheric refraction as one of their main arguments.

I might begin with another widespread misconception. Brahe's Uraniborg observatory on the island of Hven was not the first permanent observatory in Western Europe — although it may have been the grandest. Well before Brahe established Uraniborg, Wilhelm IV, Landgrave of Hesse had begun to make systematic astronomical observations and established a permanent observatory at his court in Kassel. Wilhelm was born in 1532 and died in 1592. It is clear that he had begun making observations for a new star catalog before his accession in 1567, while Brahe, still a student, was developing his interest in astronomy and acquiring his first instruments. When the pressure of official business limited his own direct involvement, the Landgrave brought assistants to his court to continue the work. The most famous were Christoph Rothmann, who served as Wilhelm's mathematicus from 1577 to 1590 and the instrument maker Joost Burghi (1552-1632), who held an official appointment from 1579 until Wilhelm's death, and then moved to Prague where he served Rudolf II, Matthias and Ferdinand II[1].

---

1. B.T. Moran, " Christoph Rothmann, the Copernican theory, and institutional and technical influences on the criticism of Aristotelian cosmology ", *Sixteenth Century Journal* , 13 (1982), 85-108.

Wilhelm established a permanent observatory on a balcony attached to his residence, an arrangement similar to the famous observing platforms used by Brahe at Hven. The Landgrave installed instruments similar to those made famous by Brahe's *Mechanica* of 1602. Many of these instruments are still preserved at the Landesmuseum at Kassel. Brahe himself visited the Kassel observatory in 1575, and conducted a long and well known correspondence with Rothmann and the Landgrave. This relationship is celebrated in a formal portrait of the Landgrave among his instruments, which shows a figure that is allegedly Brahe as a detail in the lower right corner. If this attribution is accurate it may be our only profile portrait of Tycho Brahe. More importantly, according to Gassendi, it was during the 1575 visit that the Landgrave told Brahe about the apparent retardation of the Sun's motion as it neared the horizon[2]. Hence, Brahe's attention may first have been drawn to the issue of refraction by the Landgrave and his assistants. Certainly, refraction was a major research issue at Kassel during the next decade.

Rothmann was installed at Kassel in 1577, and began an ambitious program of writing on astronomy, much of which is still preserved there. During the 1580s he wrote a Wittenberg style introduction to astronomy, a book on the comet of 1586 (sent to Brahe in draft and ultimately published by Snell in 1618) and a third book on observations of the fixed stars. The last two are particularly significant : the former repeats Pena's arguments on the substance of the heavens, but based on new observations. The latter contains an extensive treatment of refraction, and a table of corrections.

Tycho had probably begun to make observations to establish the corrections necessary for refraction during the early 1580s, in connection with his annual observations of the Sun. However, his main astronomical publication, a book on an earlier comet, was delayed from 1577 when the comet appeared to 1588, long after all the publications by other people on the same subject. The delay was probably caused by Brahe's recognition of a problem apparently unrecognized by his contemporaries. Although several people were experimenting with what is now called a Tychonic arrangement of the planetary system, only Brahe recognized that, given realistic planetary distances, the geo-heliocentric arrangement required that the spheres of Mars and the Sun intersect[3].

The solution to this difficulty may well have been influenced by Brahe's receipt of Rothmann's book on the 1586 comet. Brahe had not read Pena, who in 1557 had used arguments from optics to revive the Stoic doctrine that the substance of the heavens was a continuous fluid, the *air* or *pneuma* that for the Stoics is the source of life and intelligence. The most important argument establishing this positive conclusion, and the parallel negative conclusion that there were no Aristotelian spheres in the heavens, depended on refraction. Dur-

2. J.L.E. Dreyer, *Tycho Brahe*, New York, Dover, 1963, 80.
3. C.J. Schofield, *Tychonic and semi-Tychonic world systems*, New York, Arno Press, 1981.

ing the 1530s Gemma Frisius has used the newly introduced *radius astro-nomicus* to look for refraction as a correction to the relative distance between two fixed stars at the zenith and at the horizon. He reported finding nothing. Pena, accepting this observational result, reasoned from the absence of refraction to the absence of refraction-causing boundaries between the observer and the fixed stars. As the observer was immersed in air, he concluded that air — in the Stoic sense — must extend all the way to the fixed stars, and that here was no sphere of fire, nor any celestial orbs supporting the planets[4]. Rothmann reproduced Pena's arguments — often verbatim — in his book on the comet of 1586. When Brahe read these arguments, it would have become apparent that the celestial orbs could be merely geometrical constructions in a continuous fluid, so that the interpenetration of the spheres of Mars and the Sun no longer presented a physical difficulty[5]. Whether this conjecture is accurate, it is apparent from his correspondence with Rothmann in 1586-1587 that Brahe rapidly adopted a version of the Stoic fluid heavens, and set in motion the production of *De Mundi Aetherei Recentioribus Phaenomenis*, which appeared in 1588. The solution to the problem of the intersecting orbs was purchased at a price, if the source was Pena, for Pena had completely denied the existence of refraction. This, however, was a difficulty that Brahe's main source for Pena, Christoph Rothmann, had already surmounted.

Rothmann's *Observationum stellarum fixarum*, composed between Summer 1588 and Summer 1590 but based on earlier work at Kassel, treats refraction extensively from Chapter 15 to Chapter 22 (fols. 45v-68r). Rothmann begins with a treatment of refraction in general based on Alhazen, Vitellio and Theodosius. He goes on to consider stellar refraction as a correction to the observed position of a celestial body, in the opposite sense to parallax, asserting that it is a real phenomenon, as shown by the more accurate observations of stellar positions possible with Kassel instruments such as an astronomical quadrant. In Chapter 16 he gives a table of corrections for angles from 2 to 31 degrees above the horizon. The table contains separate entries for the Sun and the fixed stars. The two tables are identical up to 7 degrees but the correction for the fixed stars declines more rapidly than for the Sun, lagging behind more and more above 7 degrees and vanishing at 29 degrees.

In Chapter 17, Rothmann argues that stellar refraction originates near the earth, on the grounds that if the aether and the material immediately beneath the Moon differed in transparency, then refraction would be detectable all the

4. P. Barker, " Jean Pena (1528-58) and Stoic Physics in the Sixteenth Century ", in R.H. Epp (ed.), *Recovering the Stoics : Spindel Conference 1984. Southern Journal of Philosophy* (Supplement), 13 (1985), 93-107 ; P. Barker, " Stoic contributions to early modern science ", in M.J. Osler (ed.), *Epicurean and Stoic Themes in European Thought*, Cambridge, Cambridge University Press, 1991, 135-154 ; P. Barker, " The Optical Theory of Comets from Apian to Kepler ", *Physis*, 30 (1993), 1-25.

5. B.R. Goldstein, P. Barker, " The role of Rothmann in the dissolution of the celestial spheres ", *British Journal for the History of Science*, 28 (1995), 385-403.

way to the zenith. In essence, then, Rothmann continues to accept Pena's argument, and the result of Gemma Frisius on which it is based, while accommodating a limited correction for refraction near the horizon.

The next chapter introduces the main theme of Pena's refraction argument : a variety of considerations, including refraction, show that the material of the celestial spheres is not solid, carrying around embedded planets. Rather it is subtle and liquid and easily gives way to the motion of the planets. Indeed, in the next chapter Rothmann concludes, again almost in Pena's words, that the material surrounding the planets is not different from pure sublunar air. Air however may be divided into " pure " and " thick " or " dense ", the latter lying closest to the earth. It is this layer of thicker air surrounding the earth that generates refraction, and this in turn explains why, near the horizon, the distances of stars appear greater and the bodies of the same stars larger than at the zenith. In the final chapter of this section, then, Rothmann returns to the original phenomenon described in the refraction argument given by Pena following Gemma Frisius.

Brahe had clearly begun systematic observations of refraction during the 1580s, and had paid particular attention to refraction effects for the Sun between 1585 and 1589[6]. However, his main remarks on refraction all postdate his receipt of Rothmann's comet book, and the most important exchanges in his correspondence with Rothmann occur in the period after the publication of *De Mundi Aetherei* and before Rothmann's visit to Hven in 1590. This is the period when Rothmann was composing the chapters on refraction for *Observationum Stellarum Fixarum*. It is probably not accidental that Hven observations made in order to study stellar refraction are particularly numerous in 1589[7]. In his exchanges with Rothmann, Tycho quibbled about the identification of the substance of the heavens, but he accepted the general points that the substance of the heavens was fluid, that its optical properties did not change until very close to the surface of the earth, and that refraction occurred only near the horizon. Brahe proposed that refraction was caused by vapors or exhalations[8]. Although he gives a few figures in their correspondence, for example corrections to the position of the Sun, Brahe offers nothing like Rothmann's table of corrections before the *Progymasmata*, which contains tables similar to Rothmann's for the Sun and fixed stars, and adds a table for the Moon (for a comparison, see Table I). Although it is difficult to know precisely when the material assembled in the *Progymnasmata* was first produced, the printing of the book clearly postdates *De Mundi Aetherei*. Although Kepler — perhaps charitably — said that the book was written between 1582 and 1592, the first

6. J.L.E. Dreyer, *op. cit.*, 334.

7. J.L.E. Dreyer, *op. cit.*, 336.

8. Brahe to Rothmann, 17 Aug. 1588 in T. Brahe, *Opera Omnia*, 15 vols, Copenhagen, Gyldendaliana, 1913-1929, ed. by J.L.E. Dreyer, vol VI, 136-137.

chapter at least was actually written in 1588[9]. The table for the Sun also corrects the refraction value for 30 degrees from 1 minute 30 seconds (given in the correspondence with Rothmann) to 1 minute 25 seconds, suggesting that the construction of Tycho's table took place after the correspondence. Given the free exchange of work between Kassel and Hven up to that point, we must also expect that Rothmann took a copy of his *Observationum Stellarum Fixarum* with him when he visited in 1590, which would have given Tycho the complete table from Chapter 16, along with Rothmann's commentary. The picture that emerges is one of refraction work at Hven responding to and following a lead given by the workers at Kassel, and especially Rothmann, who was responsible for recording their activities in his books, and describing them to correspondents.

In his book on the fixed stars Rothmann directly connects the question of refraction with the cosmological question of the reality of the celestial spheres. This was the third round in a dialogue between Hven and Kassel that began with Rothmann's comet book, and extended through Tycho's presentation of his new system in the *De Mundi Aetherei*. In the book on the fixed stars Rothmann responded to many of Tycho's points in the correspondence as well as *De Mundi Aetherei* and included extensive discussion of arguments against the existence of celestial spheres based on refraction and comets. He also included a long chapter reconciling the new theory that the heavens were fluid with the Bible[10]. Brahe too connected the argument about refraction with the question of the reality of celestial spheres, although he insisted on an even more important role for comets in deciding the question.

Kepler, who is an heir to both the tradition that produced Rothmann and the scientific heritage of Brahe, uses the refraction argument against the reality of celestial orbs in the *Paralipomena*[11], in the *Praefatio* to the *Dioptrice*[12] where he acknowledges the contribution of Pena, and in the *Epitome astronomiae Copernicanae*[13]. The third of these is particularly interesting. Here Kepler asserts that Tycho Brahe refuted the solidity of the celestial spheres by three main arguments : first, from the motion of comets that would be obstructed by solid spheres ; second, from refraction, using arguments like those we have noted in Pena and Rothmann before Brahe, and, third, from the problem that had motivated Brahe to abandon the spheres — the intersection of the orbs of

9. J.L.E. Dreyer, *op. cit.*, n. 1 p. 186, n. 3 p. 368 ; J. Kepler, *Opera Omnia*, 8 vols, Frankfurt-Erlangen, Heyder & Zimmer, 1858-1871, ed. by C. Frisch, vol. VI, 568.

10. M.A. Granada, " Christoph Rothmann e la *teoria dell'accomodazione* ", *Rivista di Storia della Filosofia*, 51 (1996), 789-828.

11. J. Kepler, *Ad Vitellionem paralipomena, quibus astronomiae pars optica traditur,* Frankfort, C. Marnius, 1604, 112.

12. J. Kepler, *Dioptrice*, Augsburg, D. Francus, 1611, 1-2.

13. J. Kepler, *Epitome astronomiae Copernicanae*, Linz, J. Plancus, 1618-22, book IV, part I, sect. I, 439.

Mars and the Sun. When Kepler turns to the substance of the heavens, the clinching argument that the aether is less dense than air is a numerical estimate of the density of aether compared to air, again based on refraction. Throughout the careers of Rothmann, Brahe and Kepler, then, the question of the precise measurement of celestial positions and the question of substance of the heavens intertwine in the study of refraction.

Table I : Refraction of the Sun and Stars according to Rothmann and Brahe

Altitude of Sun degrees	Rothmann		Brahe		Altitude of Star degrees	Rothmann		Brahe	
	'	"	'	"		'	"	'	"
0			34	00	0			30	00
1			26	00	1			21	30
2			20	00	2	13	40	15	30
3	12	20	17	00	3	12	20	12	30
4	11	00	15	30	4	11	00	11	00
5	9	35	14	30	5	9	35	10	00
6	8	10	13	30	6	8	10	9	00
7	6	50	12	45	7	6	50	8	15
8	5	45	11	15	8	5	40	6	45
9	4	50	10	30	9	4	40	6	00
10	4	05	10	00	10	3	50	5	30
11	3	30	9	30	11	3	10	5	00
12	3	05	9	00	12	2	40	4	30
13	2	40	8	30	13	2	10	4	00
14	2	20	8	00	14	1	50	3	30
15	2	00	7	30	15	1	35	3	00
16	1	45	7	00	16	1	20	2	30
17	1	30	6	30	17	1	10	2	00
18	1	20	5	45	18	1	00	1	15
19	1	10	5	00	19		50	1	00
20	1	00	4	30	20		45	0	00
21		55	4	00	21		40		
22		50	3	30	22		35		
23		45	3	10	23		30		
24		40	2	50	24		25		
25		35	2	30	25		20		
26		30	2	15	26		15		
27		25	2	00	27		10		
28		20	1	45	28		05		
29		15	1	35	29		00		
30		10	1	25	30				
31		05	1	15	31				
32		00	1	05	32				

BIBLIOGRAPHY

P. Barker, " Jean Pena (1528-58) and Stoic Physics in the Sixteenth Century ", in R.H. Epp (ed.), *Recovering the Stoics : Spindel Conference 1984. Southern Journal of Philosophy* (Supplement), 13 (1985), 93-107.

P. Barker, " Stoic contributions to early modern science ", in M.J. Osler (ed.), *Epicurean and Stoic Themes in European Thought*, Cambridge, Cambridge University Press, 1991, 135-154.

P. Barker, " The Optical Theory of Comets from Apian to Kepler ", *Physis*, 30 (1993), 1-25.

T. Brahe, *De Mundi Aetherei Recentioribus Phaenomenis*, Hven (1588).

T. Brahe, *Astronomiae instauratae Progymnasmata*, Prague (1602a).

T. Brahe, *Astronomiae instauratae Mechanica*, Prague (1602b).

T. Brahe, (1913-29) *Opera Omnia*, Copenhagen, Gyldendaliana ; J.L.E. Dreyer, ed., 15 vols. (TBOO)

J.L.E. Dreyer, *Tycho Brahe*, New York, Dover, 1963.

B.R. Goldstein, P. Barker, " The role of Rothmann in the dissolution of the celestial spheres ", *British Journal for the History of Science*, 28 (1995), 385-403.

M.A. Granada, " Christoph Rothmann e la *teoria dell'accomodazione*", *Rivista di Storia della Filosofia*, 51 (1996), 789-828.

J. Kepler, *Ad Vitellionem paralipomena, quibus astronomiae pars optica traditur,* Frankfort, C. Marnius, 1604.

J. Kepler, *Dioptrice*, Augsburg, D. Francus, 1611.

J. Kepler, *Epitome astronomiae Copernicanae*, Linz, J. Plancus, 1618-22.

J. Kepler, *Opera Omnia.*, Frankfurt-Erlangen, Heyder & Zimmer ; C. Frisch, ed., 1858-71, 8 vols. (JKOO)

B.T. Moran, " Christoph Rothmann, the Copernican theory, and instituional and technical influences on the criticism of Aristotelian cosmology ", *Sixteenth Century Journal* , 13 (1982), 85-108.

C. Rothmann, " Christophori Rothmanni Bernburgensis, Illustrissimi Principis Guilielmi Landgravii Hassiae, etc., Mathematici observationum stellarum fixarum ", *Landesbibliothek und Muhardsche Bibliothek der Stadt Kassel. 2° MS Astron.*, 5, n° 7 (c. 1589).

C.J. Schofield, *Tychonic and semi-Tychonic world systems*, New York, Arno Press, 1981.

ACKNOWLEDGEMENTS

This research was supported in part by a grant from the Research Council of the Graduate College at the University of Oklahoma. I would like to thank Bernard R. Goldstein and Gérard Simon for help and criticism.

Hieronymus Fabricius d'Aquapendente : *De visione*, Ending of the Perspectivistic Tradition

Erwin De Nil - Mark De Mey

Introduction

Can a failure be as instructive as a hit ? To understand the scientific mind it might be more instructive to analyse the research path of a competent scientist losing out from outsiders or neophytes than reconstructing the success story of the discovery. In this century, such might be the case with Linus Pauling's failure to reveal the structure of DNA. While being among the greatest authorities in the field and renowned contributor, quite competent and with the best chances to make the major breakthrough, he lost out against the younger team of Watson and Crick[1]. The insider failed, despite his wealth of knowledge. The outsider tackled the issue, despite a new disguise, head on and won. At the beginning of the seventeenth century, a comparable situation seems to apply to optical research elaborated by Fabricius D'Aquapendente and Kepler.

It is rather conventional, although somewhat simplified, to compare the mechanism of our human eye to a *camera obscura*. As a result of the different refracting humors of the eye, a sharp but " inverted " image of the outside world is said to be projected upon the retinal layer or retina (Fig. 1). Looking for the roots of this current " retinal image model " one has to go back as far as the year 1604 when Johannes Kepler (1571-1630) published his volume *Astronomiae pars optica*, entitled somewhat remarkably as *Ad Vitellionem Paralipomena* (Fig. 2).

At first sight it was formulated as some " minor additions " to the work of Witelo, the Polish author of one of the most widespread optical works during the Middle Ages. Diametrically opposed to these authors of the medieval *perspectiva* studies, Kepler not only abandoned in his publication their conviction that an upright image should be maintained within the eye but moreover, he

---

1. See e.g. T. Hager, *The Life of Linus Pauling*, 1995.

integrated the *camera obscura* model (Fig. 3) known from astronomy into the study of the eye. He set out in an unambiguous way the full geometrization of light and the mechanism of the eye as a pure *dioptrical* medium.

It has been ascertained that Kepler, to point out the precise path and ending of each light ray, arrived at his discovery by implementing a rather popular drawing technique of renaissance painters. In fact, Stephen Straker has demonstrated quite convincingly that Kepler's method was based upon Dürer's who traced the trajectory of light rays by means of ropes to depict images in a consistent perspective way[2] (Fig. 4). It was due to the popularization made afterwards by scholars such as Plempius, Kircher (Fig. 5) and, most of all, Descartes (Fig. 6), this model of the eye became consolidated as standard view on the way we think our eyes are functioning.

However, such an idea was far from being common in the past. Immediately anterior to the retinal image model of Kepler, one comes upon the rather neglected Latin treatise *De Visione*, also referred to as *De oculo visus organo* (Fig. 7) and written by the Italian physician Hieronymus Fabricius D'Aquapendente (1533-1619). Published in December 1600 in Venice, only four years before Kepler's edition, his account looks, confronted with Kepler's, at first view, almost archaic packed up as it is with antique reminiscences. Ensuing the main stream of antique and Arabian as well as medieval studies on the eye, Fabricius still upholds to an upright image. He does not mention at that time the already-known comparison between the camera and the eye.

In stead of the retina, he still insists on another part of the eye, namely the *lens* — or following the terminology in use at that time — the *crystallinus humor*, the *crystalline humor*, as the principal and sensitive part of the eye[3]. As a result, Fabricius, being an eminent professor for more than 35 years at the University of Padua, one of the leading medical centers of Europe, is one of the latest representatives of the old so-called " perspectivistic " tradition.

At the age of 67 years Fabricius, professor both in anatomy and surgery, finally decided to assemble the results of his lifelong research. In that very first work of him, *De Visione*, we get the core of what he called " the wonderful artwork of the eyes "[4]. Like all other works that followed, he conceptualized his ophthalmologic work, considering carefully in a sequential fashion each of the different parts of the organ, from a strong " functional " outlook[5].

After an extensive description of each of the different parts of the eye, always including their etymological history[6], he broadly analyzes and criti-

2. S.M. Straker, *Kepler's Optics : a Study in the Development of Seventeenth-Century Natural Philosophy*, Indiana University, Ph.D., 1971.

3. *Cf.* Part I, ch. 7 : *Crystallinum esse visionis auctorem, evidentissima demonstratione probatur.*

4. *Cf.* the praefatio : *qui oculorum admirabile tractat artificium.*

5. *Cf.* Part III : *De utilitatibus tum totius oculi tum partium ipsius.*

6. *Cf.* Part I : *Partis primae de oculi dissecti historia.*

cizes in a rather redundant and verbose style the various statements concerning the mechanism of the visual system, firmly based upon his own observations and guided by the claims of antique and Arabian authorities. Part II e.g. is dedicated entirely to the philosophical theories on the *modus visionis* proposed by the antique philosophers, Aristotle, Plato, the Epicurean and Stoic writers. Hence he demonstrates all his encyclopedic knowledge, the beginnings of the geometrical optics as well as the anatomical findings of others but also his own personal extremely fine observational registrations.

Grosso modo the text *De Visione* deals with the anatomical and physiological aspects of the eye. Being the first section of a triple volume entitled *De Visione, voce et auditu*, his mainly medical perspective on the functioning of the eye is integrated into a more comprehensive study of sense perception.

PROGRESS IN ANATOMICAL REPRESENTATION

In addition to the textual description Fabricius inserted very promising numerous real sized illustrations of the dissected eye parts (*cf.* Fig. 8, 9, 10, 11). As he claims himself explicitly in the preface, these drawings are intended to challenge the famous woodcuts of one of his most famous precursors at Padua, Andreas Vesalius. In fact, on the field of ophthalmological iconography, Fabricius distinguishes himself explicitly in confront to Vesalius' eye representations (Fig. 12) for his additional images, the real sized format, his use of color and a double layout (an aristocratic hardback volume and a paperback version for student purposes).

The supplementary pictures might have come forth out of his concern to unroll different perspectives to do more justice to some eye components. This altering viewpoint and depth effect in some of his representations, to say his attempt towards a three dimensional depiction remarkably runs parallel to Fabricius' interest for definite motions aspects in the visual process. With respect to Vesalius we are confronted several times with a shift in the attention. Fabricius clearly insists on those aspects or phenomena that clarify the *actio* and physiology of the visual perception (Fig. 8, 9, 10, 11).

Closely connected to this predilection for a full iconographic representation, one has to refer to Fabricius' engagement to start the construction of an *anatomical theater* in 1594 (Fig. 13)[7]. The visual demonstration of autopsy for a broad audience goes back to the same concern to present a precise and detailed observation of the subdivisions of the body to a large public. This attitude of Fabricius is characteristic for the increasing influence for " visualization " while the research method itself will gain more and more weight in the medical discipline.

---

7. In his article Rippa Bonati, 1988, 163, even makes a comparison between the shape of the theatre and the depicted shape of the lens in the *De Visione*.

ORDINARY OPTICAL KNOWLEDGE

In contrast to these extremely " accurate anatomical drawings " of the eye parts are the rather " roughly made geometrical schemes " in the third part of the book. One might speak of a discrepancy between his very respectable anatomical knowledge and the comparatively speaking poorly elaborated geometrical drawings.

In general the latter illustrations handle with basic optical and geometrical postulates that intend to demonstrate the validity of popular perceptual theorems one can easily verify. Most of these theoretical claims are borrowed from former ancient and medieval optical authors (Bacon, Witelo, Pecham…), some of them were already known to Euclid. Within these strongly schematized representations incorporating the letter indications and concepts, we can identify as followed :

1) " The visual pyramid as an illustration of the relation between distance and the width of the visual pyramid " (Fig. 14). Here Fabricius recurs directly to notions made already in the geometry of Euclid and in the refined experimental optics of Ptolemy. By means of a visual pyramid seen as bundle of radiation, with the apex in the eye, the perception of size in relation to the distance of objects is established.

2) " A notion of binocular disparity : the different range of one, resp. two eyes " (Fig. 15). This observational precision was not so well-known : e.g. the geometrically experienced I. Danti[8] (Fig. 16) strangely holds up the mistake that both eyes cover the same visual field.

3) " The position of the eyes with respect to the most appropriate sight field " : the disparity of the eye image and their role for stereopsis or depth perception is not acknowledged (Fig. 17).

4) " The benefit of a spherical shape of the eye with regard to the maximal visibility " (Fig. 18).

5) " The refraction of light rays in a different medium " (Fig. 19 & 20).

6) " The spherical shape of the cornea " (Fig. 21).

7) " The parallelogram : an illustration of a theorem of Euclid's Elements ", (XI.6) and " the reduced maximal visibility in case of a flat eye surface " (Fig. 22 & 23).

8) " The convergence of perpendicular rays into a spherical shape " following the proposition of Alhazen I, 33 & 34.

9) " The shape of the light describing a pyramid of rays with the pupil as top ".

8. Danti, *op. cit.*, XXX : *Quella cosa che da noi è veduta con amendue gli occhi, ci apparisce una sola, & non due, perche le piramidi, che nell'uno & nel l'altro occhio dalla cosa veduta vengono a formarsi, come sono le piramidi che vengono alli due occhi E, F hanno la medesima basa, & l'assi dell'una & dell' altra piramide che vanno a gl'occhi, escono dal medesimo punto G, & perciò tanto vede un' occhio, come l'altro.*

10) " The surface of the uvea larger than the cornea ; the necessity of refraction of the rays in order to allow incidental light rays entering into the eye ". Correlation between angle of diffraction, the latitude of the cornea surface and the width of the pupil opening.

11) " The different refractions from and towards the perpendicular in the aqueous humor ".

12) " The different refractions in the eye to demonstrate precisely the convergence of the incoming light rays towards the center of the crystalline humor ".

13) " The flattened spherical surface of the cristalline : in contrast with a circular crystalline : a greater part of the circumference can be in contact with the rays coming from the objects ".

14) " The exact location of the centra of the eye, the aranea and the cornea ".

15) " The rectilinearity of light rays ".

16) " Distinct and peripheral vision and maximal range of visibility in case of a fixed eye position, resp. with moving eye and head " (Fig. 24).

17) " The parallel pattern of the movement of both eyes ".

The disconnected character of these optical issues from his anatomical and philosophical statements may be partly determined by the education programme he got. To repeat, Fabricius was a typical product of the Faculty of Arts in Padua. After leaving his home place Aquapendente in the midst of Italy, he went to the northern city and built up his whole career at that renowned University. The normal curriculum there involved medicine and philological courses taught together within the same faculty. The optical and geometrical aspects in the medical circle were treated as of minor importance. Therefore the shortage of a more geometrical and mathematical oriented education can account for some but not all of his misunderstandings.

However Fabricius as well made attempts to go beyond his familiar research domain. We possess an interesting autograph of Galileo illustrating that Fabricius undertook an interdisciplinary *démarche* that was not prevalent. From Galileo's defense it can be inferred that the interest of Fabricius into a speciality that was not his own, clearly was the envy of a colleague who strictly held on to the traditional medical practice.

However the mathematical way to approach the problems was not completely unfamiliar to doctors of medicine that were educated in Padua. As a counterexample one can refer to the later on much celebrated pupil of Fabricius, William Harvey. Partly due to Fabricius' discovery of the valves in the veins, Harvey arrived at his revolutionary innovation of the blood circulation but in order to lay open the misconception he really did proceed to an actual " quantification " of the problem. When Fabricius is confronted with his crucial morphological data, he holds on in its comment, in sharp contrast with Harvey, to a strictly " qualitative " description of the given facts, leaving over

the quantitative interpretation to others after him : *Habebunt enim curiosi inda-
gatores operum naturae, ubi multa contemplari possint* (Part III, 8) (Fig. 25).

FABRICIUS' MODUS VISIONIS

Let us briefly summarize his optical account. In his philosophical discussion[9] of the process of visual perception Fabricius makes visible how greatly
he was indebted to the antique tradition. The activity of the eye, the *modus
visionis*, is organized around the sensation, capturing the external aspects, i.e.
the visible appearances of the outer surface of the bodies. All these aspects are
reducible to unspecified *passiones*, influences, impressions or affections of the
light. Recognition of all the visual objects can be achieved solely by touch or
contact between the perceptible entities and the nerve that comes out of the
brains where the cognitive competence is located.

This concept corresponds to the Greek's standing conviction that all sense
perception is realized by means of a sort of contact. The workable possibilities
then are rather limited. Still Fabricius gives a survey of all hypotheses known
to him : the extramission theory proposed by the " Stoics " with the emanating
πνεῦμα, the antagonistic intromission idea of the " Epicureans ". After that he
overlooks as a third alternative the halfhearted solution rendering the connection half way attributed to Galen, further on the extramission idea of Plato and
others who hold on to a radiating inner fire.

In Fabricius' opinion the only legitimate explanation is supplied by the
remaining authority of Antiquity, Aristotle. Seeing is realised by means of a
medium. This intermediate substance becomes changed or transformed by the
external qualities of the objects. To see, in other words, is " a kind of affection,
to undergo some influence by something ". Nature has arranged it so that there
is something that itself changes under the influence of the objects. The image
or representation of the object then is light that has been modified under the
influence of the object. The light, " lux ", operates also as a means of transpor-
tation for the image. The substance for the light is the diaphanous, τὸ
διαφανές or the transparent. Without this diaphanous " go between " no light
can spread. Moreover, the sense organ is susceptible for the form, the
"*forma* ", without the substance. With the exception of touch, senses don't
need any substance : they merely modify the corpus or the structure of the
object. Only the diaphanous that represents the body, will undergo some alteration.

Therefore two principles lay at the foundation of seeing, as the fundamen-
tals for the transmission of visual forms : light and the diaphanous. The definition and the significance of these notions go back to Aristotle. His
interpretation of light, although not perfect according to Fabricius, defines light

---

9. Part II, *De actione oculorum.*

as a quality or property that assimilates colour and form[10]. The combination of both working principles and thus the mechanism of the visual process takes place according to an ancient mechanism to encountered already in Empedocles writings : the *similia similibus* principle. Ultimately this dogma " similar things correspond to each other " ends up to be the determining criterion also when one has to choose out of the candidates for the central organ in the eye. The analogy between micro- and macrocosmos has been maintained. Understanding, i.e., cognition and perception, is to take place when similar things are matching analogous. The sense organ must be aligned and has to be strictly matched with the substance that sets the sense organ into motion.

POSITION, SHAPE AND FUNCTION OF THE LENS

A critical issue in the ophthalmological literature concerned the search for the most prominent part of the eye. To be in line with the Greek and Arab predilection for the crystalline humor or lens, Fabricius set up a specific logic on his own, weighing in detail the pros and cons for each of the proposed candidates. One can admire his logic : for Fabricius the reception of light impressions, as postulated before, can only be realized through a diaphanous transparent medium to arrive at the optical nerve. Ergo, one of the three transparent bodies present in the eye — the vitreous, the aqueous or the crystalline humor — has to be the central part. By exclusion he comes to explain why the crystalline humor and only that body is eligible for that main function and not the vitreous or the aqueous humor. However this deadly choice is dictated by his erroneous premise. What follows is a lengthy argumentation to defend his prejudice[11].

By this assumption Fabricius is confronted with two problems. First, accepting that the visual, so to speak, " crystallises " in the crystalline body, how then is the " connection to the optical nerve " accomplished where, according to Fabricius, data of vision fuse with the cognitive capacity coming from the brain ? He refers to Witelo, assuming a connection between the crystalline and the retina. This connection was later on taken over by Fabricius' pupil Jessen.

Kepler in his *Paralipomena* for his part used Jessenius as a direct source of information for his anatomical knowledge. Kepler however, for obvious reasons, will prefer the suggestion proposed by the Swiss physician Felix Platter

---

10. Part II, ch. 3 : *Etenim ut scribit Aristoteles, libro de Sensu & Sensili per id quod est esse aliquid, lumen est, sed non motus aliquis, quando diversa ratio est in alteratione & latione... lux vero incorporea sit, & simul tota, ac subito (ut dicit Aristoteles loco citato) & in instanti moveatur : non propterea damnandus est hic loquendi modus, cum potiorem non habeamus... Quoniam vero lux qualitas quaedam est, quae nisi & ipsa subiectum, seu materiam habeat, neque consistere, neque vehi ullo pacto potest...*
11. *De visione*, Part II, ch. 6 & 7.

(1536-1614). In his *De corporis humani structura et usu libri III* from 1583, Platter did not consider such connection to be indispensable because he situated the visual recognition of the object on the retina itself. Fabricius however does not seem to have known the work of Platter.

Problem two regards the function of the transparent " vitreous body ". When one has chosen for the lens, what kind of role is left for this transparent medium since all the entering light rays are already bundled together in the crystalline humour ? Fabricius seems to be aware of that difficulty. " Yet for which reason the light still has to permeate the vitreous humour, that has not been found yet "[12]. Anyhow he would have been familiarised with the ideas of optical writers who proclaimed that the visual process was not concluded in the crystalline. The explanation may seem for us rather obvious, Fabricius however tried by all means to solve this puzzle that seemed very contra intuitive to him. He ends up by inventing several responses, the most effective being that the vitreous humour must prevent the adverse reflection towards the crystalline body.

And what about the function of the retina so crucial for the visual process in Kepler's theory ? For Fabricius this layer is confined to a merely supportive, nutritive function. In part III, chapter VII he even points out that the retina could be a negative influencing factor because of the fact that its thickness might reflect the light and even double and strengthen it. As a result, one could say that Fabricius considers the retina a kind of concave reflecting mirror and certainly not as a receptive organ or screen[13].

Concomitant with the attribution of a central role, is the position of the lens. Fabricius is stubbornly sticked to the persisting but false idea that supposed the crystalline body to be in the central position of the eye bulbus[14]. He is even willing to overrule his personal observations. In the illustration we can witness how he neatly indicates the correct position, closely to the outer surface of the bulbus (Fig. 25). This textual adherence to the tradition of a central position while his accompanying illustration is revealing the anatomically more correct

---

12. *De visione*, Part III, ch. X : *at cuius rei gratia lucem per vitreum transire oporteat, id sane compertum nondum est, praesertim cum potius contrarium credere consentaneum sit.*

13. *Cf.* Part III, ch. 7 : *...ne lux ultra Crystallinum reflectatur, ea scilicet specie affecta, quam ex retinae contactu contraxisset, idque sine ulla sensus videndi, aut animalis utilitate.*

14. Fabricius' conflict between his own differentiated observational registrations and the authoritative claims, as well as his attempt to reconcile both opposing views, emerges esp. in Part I, ch. 7 : *In medio oculi ponitur crystallinus praecipue in hominibus : at quadrupedibus & iis quae prona incedunt inferius spectare oculique infernae orbitae vicinius esse videtur. Quod si illud iam quaeratur, in anterioremne, an posteriorem partem magis propendeat crystallinus, puto in anteriorem magis prominere : quamvis si quis vitrei altitudinem, cum corneae extuberantia, & aquei humoris copia & crystallini in vitreum immersione conferat, facile positionem aequalem esse iudicabit ex eo igitur quod in medio est...* His final submission to the stock opinion becomes evident from Part III, ch. 7 : *Neque insuper est Crystallini positio omittenda, quae merito medium locum sortita est, non tantum quod praecipuae, & principi oculorum parti medius locus merito debetur quo ab omnibus externis iniuriis sit remotissimus... Quare necessario in medio sibi ministrandum fuit : ...Quod vero in medio sit positus...*

frontal position is all the more surprising when compared a predecessor such as Vesalius (see esp. first eyediagram on Fig. 12).

As a supplementary difficulty to the correct interpretation of the eye it may be stressed that the *dissection technique* may have played an important role. Due to his particular position the lens got easily moved backwards and so the practitioner could have the impression that the lens was situated in the center of the eye just as Galen and others had reported.

The displacement could be avoided by making use of a frozen eye, but he does not mention about that. Anyhow, Kepler gave no thought to the dissection and their problems, he just took over the results of renowned men of his time including Fabricius', and so he was not hindered to develop the right insight in the optical functioning of the eye. The physicians however remained for a long time under the long-lasting illusion that the lens was located in the middle. The correct interpretation of the front position, though correctly distinguished by some antique writers, became lost later on and was preserved until by the late 1550s the misapprehension was rediscovered again. It is hard to find out who did the primary finding. It might have been the Spanish Juan Valverde (Fig. 26) or Colombo, both however are accused from plagiarizing heavily Vesalius' *De fabrica*. Vesalius himself does not mention the correct position in his *Fabrica* but in a posthumous treatise of him, the *Anatomicarum Gabrielis Falloppii observationum examen (1564)*, an examination on the newest publication of his friend and follower Gabriele Falloppius of Padua, he admits while referring to a new dissection technique, his former error on the position of the lens[15].

Another issue apparently important at that time concerned the shape of the lens. Based upon an external and common likeness, antique authors like Rufus of Ephese had denominated the lens as φαχοειδής, i.e. they compared it to a φαχός the seed of a lentil plant. Translated into Latin the lenticularis or lenti-form *lentis* gave its name later on to the *lens*. Fabricius eager to indulge into philological parentheses demonstrates repeatedly full familiarity with the many terminological variants (e.g. *glacialis, crystallinus, gutta humoris, grando*) for the crystalline body. As it was noticed already by Galen and Arab authors after him (e.g. Honayn Ibn Ishaq), Fabricius was equally well aware of the fact that the crystalline body actually contains a somewhat dyssymmetrical shape with the anterior part flattened and the posterior part more convex.

Although detailed anatomical analyses of the different parts of the eye were available to the physicians at the University of Padua, they had less clear ideas concerning the effective use of this peculiar shape. The optical function of the lentiform element was only understood in global sense. Take Vesalius, who,

15. Vesalius, *Falloppii examen*, 161-163 : *cristallinum humorem magis quam opportuit, in posteriora retrusi... In huius vitrei humoris superficiei medio, cristallinus ita reponitur, ...ortum anterius, quam hactenus feci esse statuendum, ...est tamen is humor ita retrorsum collocandus, ut exacte oculi centrum ipsi tribuamus.*

while working some decennia before Fabricius in Padua, only dared to give a provisional description in his *De Fabrica*[16]. According to Vesalius, one might conceive the transparent and lucent crystalline body as a kind of magnifying glass made out of two small biconvex glass disks put together. The *specilla* Vesalius referred to, in this instance meant no longer mirrors but designated genuine lenses. In fact we are confronted here with one of the very early descriptions of a yet non-specific comparison between the anatomical lens and the artificial elaborated glassy segments.

The global understanding of these magnifying glasses were already well known before Vesalius. Several sources that go back more than 300 years before him as e.g. Roger Bacon had reported on their advantageous use for bad sighted elderly men. Skilled craftsmen active in glass industry especially in Italy but soon also widespread over Europe possessed the technological know how, required for making these appropriate lenses and loupes. We can rely on it since at least the end of the 13th century, representations of spectacles appear in manuscripts, upon frescoes and paintings. Recently Neamann hinted at a very early example that, although somewhat debatable, was found in a Ghent Psalter dating from 1240-1270, showing a strange *drôlerie* of a bird dragon wearing spectacles (Fig. 27 and fig. 28).

With regard to the *specilla* or spectacles, Fabricius only briefly touches upon them in order to show their refractive function in the sequence of alternating denser and rarer refractive media[17]. Fabricius is obviously well aware of their dioptrical function. However, the strict parallel between spectacles and the crystalline humor is not shown here as in Vesalius. At another location in *De Visione*, the correspondence between the eye lens and a glassy correlate is explicitly mentioned to account for the different refractive humors, arguing that the bundle of entering light rays come together immediately behind the crystalline body whereas in the case of a piece of glass the point of conver-

---

16. *De Fabrica*, Book VII : *Est enim is humor, optimi crystalli instar pellucidisimus, & omnia quibus iam exemptus, vitri alicuius modo imponitur, impensae quorundam utrinque extuberantium specillorum ritu adauget. Ea autem est consistentia, ut a caeteris oculi partibus per sectionem liberatus, humoris liquoris ve cuiusdam modo non diffluat : sed molliusculae instar cerae, suam asservet formam. quae etsi rotunda appareat, non tamen globum adamussim rotundum refert : sed globi speciem exprimit, qui anteriori & posteriori parte non secus comprimeretur, quam si ex lignei globi penitus exacteque rotundi medio, secundum aequidistantes lineas orbem crasiusculum serra exemisses, & dein duas globi partes denuo conglutinasses : adeo ut praesens humor in anteriora & posteriora minus protuberet, quam secundum latera, & superius & inferius, quodammodo scilicet ad lentis similitudine unde etiam hunc humorem " φακοειδής " Graecis nuncupatum fuisse arbitror.*

17. *De Visione*, Part III, V : *Atque ut summatim dicam, si quod in specillis ocularibus evenit, huc adferatur ; huius rei plena notitia, ni fallor, habebitur. Ocularia specilla, sive katoptra ut Graeci loquuntur, duo potissimum praestant. Visilia enim quae obscure viderentur, & clariora, & maiora faciunt, quod evenit, quia primum lux ab aere, quod est rarius diaphanum, intrat in ocularia specilla, quae sunt crassiori diaphano, & inde in aerem rursus pervenit rarius diaphanum, a quo in Corneam densius, & a Cornea deinde in rarius, videlicet humorem aqueum, & ab hoc tandem in densius diaphanum, nimirum humorem Crystallinum.*

gence happens somewhat farther[18]. So in confront with Vesalius' merely general metaphorical description of biconvex lenses, Fabricius attempted a more elaborated optical exposition of the function of the crystalline body. One would have wished Fabricius, being so near to the solution, had experimented somewhat more with lenses. Though, as mentioned before, such experiments would cast serious doubt upon the central role of the crystalline body because one then would have to locate the act of seeing not in the lens but behind it. Can this be the reason why he was not interested or did not want to calculate the further trajectory of these light rays behind the glass ?

As a result of the light convergence a scrupulous optical account would have troubled and disorganized thoughtfully the upright form of the images in the eye. So any longer perpetuation of the light trajectory was out of his concern since, as we stated before, Fabricius the receptive and cognitive capacity was already attributed to the lens. What came behind, was neither effective nor useful. On the contrary, it was rather an embarrassing obstacle for his theory[19].

Therefore the compulsory requirement of the central role of the crystalline body impedes Fabricius even from considering other alternatives such as the vitreous body or the retina, mostly because he was unwilling to give up the strong belief of an upright image in the eye. Only extraordinary open minded and uncommitted thinkers like Kepler were to overcome the bias that hampered the optical writers.

CONCLUSIONS

In sum, Fabricius' life and work was deemed to reside within a Galenistic and Aristotelian tradition that no longer is accepted but still was alive at the time Kepler formulated the current view. Fabricius was the leading authority on anatomy, writing a major volume on the structure of the eye, but failed to discover the retinal image that constitutes the core of the new theory of vision for which Kepler surprisingly laid down the foundation. Kepler was the outsider, tackling the issue of vision as an astronomer, interested to learn to what degree he could rely on the accuracy of his visual perception. The implicit neg-

18. *De Visione*, Part III, 10 : *alterum vero postea dispergi & evanescere uti videtis se habere post phialam aqua plenam luci oppositam : in qua lux primum unitur & clarior redditur, postea dispergitur, & evanescit. Sed neque etiam in hoc connivere, aut inops consilii esse natura visa est, cum providerit, lucis hanc unionem prope Crystallinum in vitreo fieri non longius ab ipso contingere, ne dum posteriores attingere partes : quod factum puto, per posticam Crystallini rotunditatem, seu extuberantiam, id quod & Alhazen innuit, & omnino ita esse compertum habemus, si Crystallinum, & vitrum (inquam) ex altera parte anteriorem Crystallini rotunditatem aemulans, ex altera planum factum, luci opponas. liquido enim apparebit, lucis unionem prope Crystallinum accidere, in vitro vero longius contingere, & quod maximum est, confestim unitam lucem obumbrari neque ulterius in vitreo progredi, quae omnia multo exactius apparent in lucii piscis Crystallino, & vitreo luci opposito, ob eiusdem exactam rotunditatem, atque haec quarta causa est, ob quam reflexio lucis ad Crystallinum prohibetur.*

19. *De Visione*, Part II, 7 : *Ergo in crystallino duntaxat fit tum receptio, tum dignotio.*

ative connotation on the labor of all those forerunners like Fabricius seems almost inevitable.

All in all the *De Visione* of Fabricius presents a composite, yet in some sense contradictory character. Apart from the preponderant close attention upon the classics (esp. Galen, Aristotle) and the medieval Arab authors like Alhazen, Fabricius has given initial impetus towards a physiological exposition that tried to break with the traditional dogma's of the *perspectiva*. His descriptions are a sign of conscientious observation and a meticulous registration and illustration of his research object. To some extent his methodological choice, borrowed from Galen, namely to treat always separately the different functions of the particular parts, induced at the same time the reason for its failure. The latest breakthroughs in the domains of physiology and optical functioning of the eye detected by contemporaries as Della Porta, Maurolyco, and Platter did not reach him or got only a partial assimilation in his *De Visione*. Hence an effective and thoughtful understanding into the entire optical process of rays within the eye was missing.

His persistence to hold on to the antique dogma of the lens as the primary and sensitive organ of visual perception obstructed a consequent continuation of the trajectory of the light rays in the eye. Kepler who took that as its personal challenge, was able to visualize this extension in a unequivocal way and he succeeded, immediately after him. Fabricius' *De Visione* on the other hand remained caught within the finalistic Aristotelian philosophy of nature.

This frame of reference though questioned on certain topics he preferred to maintain. That system made out his education, his schooling, his books he worked with, he referred to. Therefore his works may appear somewhat ambiguous. This might not surprise too much since Fabricius finds himself exactly in the midst of a split, a rupture where the Aristotelian scientific network crumbled away in favor of a new philosophy of nature. In that way Fabricius could represent, could be the incarnation of a dichotomy that was rising up as good as everywhere at that time between scientific systems of knowledge that were based upon the principle of the *auctoritas* and the modern experimental methodology.

SELECTED BIBLIOGRAPHY

H.B. Adelmann, *The Embryological Treatises of Hieronymus Fabricius of Aquapendente*, vol. II, Ithaca, New York, Cornel University Press, 1942, (ed. 1967).

Aristoteles, *The Complete Works of Aristotle*, J. Barnes (ed.), Princeton, Princeton University Press, 1984.

Atti della Nazione Germanica Artista nello Studio di Padova, Monumenti Storici pubblicati dalla Regia Deputazione di Storia Patria, Venice, A. Favaro (ed.), 1911.

Atti del XX congresso nazionale di storia della medicina, Roma, 10-11 october 1964, Società italiana di storia della medicina.

D. Barbaro, *La pratica della perspettiva... Opera multo Profittevole a Pittore, Scultori, et Architetti*, Venezia, 1568 (facsimile).

J. Barozzi da Vignola, *Le due regole della prospettiva pratica*, Roma, I. Danti, 1583 (facsimile).

J.I. Beare, *Greek Theories of Elementary Cognition from Alcmaeon to Aristotle*, Oxford, Clarendon Press, 1906.

J.J. Bylebyl (ed.), *William Harvey and his Age. The Professional and Social Context of the Discovery of the Circulation*, Baltimore, London, The Johns Hopkins University Press, 1979.

J.J. Bylebyl, *The School of Padua : humanistic medecine in the sixteenth century*, 335-370.

A.C. Crombie, " Kepler : De modo visionis ". A translation from the Latin of " Ad Vitellionem Paralipomena ", V, 2 and related passages on the formation of the retinal image, in : *L'aventure de la Science : Mélanges Alexandre Koyré*, vol. I, Paris, Hermann, 1964, 135-172.

A.C. Crombie, *Styles of Scientific Thinking in the European Tradition*, London, Duckworth, 1994.

A. Cunningham, " Fabricius and the " Aristotle project " in anatomical teaching and research at Padua ", in A. Wear, R.K. French & I.M. Lonie (eds), *The medical renaissance of the sixteenth century*, Cambridge, Cambridge University Press, 1985, 195-222.

G. De Bertolis and C. Agostini, *Anatomia e fisiologia del cristallino nel loro svolgimento storico*, Acta Medicae Historiae Patavina, vol. IV, 1957-1958, 1-34.

R. Descartes, *Traité de l'homme*, Œuvres de Descartes, Ch. Adam & P. Tannery, Paris, 1909.

E.J. Dijksterhuis, *De mechanisering van het wereldbeeld*, Amsterdam, Meulenhoff, 2nd ed., 1950.

Hieronymus Fabricius dí Aquapendente, *De Visione, Voce et Auditu*, Venetiis, per Franciscum Bolzettam, 1600.

Fabricius dí Aquapendente, *Opera omnia, Anatomica et Physiologica*, Leiden, Van Kerckhem, 1738.

A. Favaro, " L'insegnamento di Fabrizio d'Aquapendente ", *Monografie storiche sullo studio di Padova. Contributo del R. Istituto Veneto di scienze, lettere ed arti alla celebrazzione del VII centenario della università*, Venice, 1922, 107-36.

A. Favaro, *Galileo Galilei e lo Studio di Padova*, Padova, Editrice Antenore, 1883, (reprint 1965).

G. Favaro, *Girolamo Fabrici d'Aquapendente e la medicina Pratica*, Estratto dal Bollettino Storico Italiano dell'Arte Sanitaria, XXVI, n.1., Roma, 1927.

G. Favaro, " Contributi alla biografia di Girolamo Fabrici di Acquapendente ", *Memorie e documenti per la storia della Universita di Padua*, Padua, 1922, 241-348.

G. Federici Vescovini, *Studi sulla prospettiva medievale*, Torino, G. Giappichelli, 1965, anast. reprint 1987.

Graefe-Saemich, *Handbuch der Gesamten Augenheilkunde*, Leipzig, Hirschberg (ed.), J. Von Wilhelm Engelmann, 1899-1918.

M.D. Grmek, " La sperimentazione biologica quantitativa nell'Antichità ", *La vita e le forme i numeri, biologica*, I, Ancona, 1988, 1-25.

J. Kepler, *Les fondements de l'optique moderne : Paralipomènes à Vitellion* (1604), (translated and annotated by Catherine Chevalley), Paris, Vrin, 1980.

J. Kepler, *Gesammelte Werke*, (ed. C.H. Beck), München, 1939, including : *Ad Vitellionem paralipomena quibus astronomiae pars optica traditur potissimum de artificiosa observatione et aestimatione diametrororum deliquiorumque Solis et Luna*, 119-398.

A. Kircher, *Ars Magna Lucis et Umbrae In decem Libros digesta...*, Rome, Hermanni Scheus, 1646.

H.M. Koelbing, *Renaissance der Augenheilkunde 1540-1630*, Bern, Hans Huber, 1967.

T.S. Kuhn, *The Structure of Scientific Revolutions*, Chicago, The University of Chicago Press, 1962.

D.C. Lindberg, " The Genesis of Kepler's Theory of Light : Metaphysics from Plotinus to Kepler ", *Osiris*, 2 (1986), 5-42.

G. Lise, *Fabrizio d'Acquapendente*, Acquapendente, La Commerciale, 1988.

J.S. Neaman, " The mystery of the Ghent bird and the invention of spectacles ",*Viator*, vol. 24, Berkeley (L.A.), London, University of California Press, 1993, 189-214.

F. Perazzi and A.M. Perzazzi, " Considerazioni sull'applicazione del metodo sperimentale in medicina da parte di Galileo Galilei ", In *Atti del XX congresso nazionale di storia della medicina*, Roma, 10-11 october 1964, Società italiana di storia della medicina, 65-71.

J. Pergoot, L. Thys & E. Van Derstappen, *Fysica, voor studierichtingen met beperkt leerplan*, deel II, Brugge, De Garve, 1980.

T. Pesenti, " Galenismo e 'novatio'. La scuola medica vicentina e lo studio di Padova durante il periodo veneto di Galileo (1592-1610) ", *Medicina e biologia nella rivoluzione scientifica*, (red. Lino Conti), Porziuncola Università degli studi di Perugia, (series Ricerche filosofiche), s.d., 107-147.

Vopiscus Fortunatus Plempius, *Ophthalmolographia sive Tractatio de oculo*, Leuven, 1632, 2nd ed., 1647.

L. Premuda, *Storia della cultura veneta. La medicina e l'organizzazione sanitaria*, 1984.

L. Premuda, *Storia della iconografia anatomica*, Milano, 1957.

J.H. Randall Jr, " The Development of Scientific Method in the School of Padua ", *Journal for the History of Ideas*, 1940, 177-206.

M. Rippa Bonati, *Le tradizioni relative al teatro anatomico dell'Università di Padova con particolare riguardo al progetto attribuito a Fraî Paolo Sarpi*, Acta Medicae Historiae Patavina, 35-36, 1988, 145-168.

V. Ronchi, *The Nature of Light. An Historical Survey*, translated by V. Barocas, London, Heinemann, 1970.

R. Scipio, *Girolamo Fabrici d'Acquapendente*, Viterbo, Agnesotti, 1978.

R.E. Siegel, *Galen. On Sense Perception*, Basel, S. Karger, 1970.

G. Simon, *Le regard, l'être et l'apparence dans l'optique de l'antiquité*, Paris, Seuil, 1988.

C.J. Singer, *From magic to science*, New York, 1928.

N.G. Siraisi, *Arts and Sciences at Padua. The Studium of Padua before 1350*, Toronto, Universa, 1973.

U. Stefanutti, " Le pitture dell'anatomia di Girolamo Fabrici d'Acquapendente ", *Rassegna Medica*, n. 1 and 2, 1957.

S.M. Straker, *Kepler's Optics : a Study in the Development of Seventeenth-Century Natural Philosophy*, Michigan, Ann Arbor, 1971.

O. Temkin, *Galenism, Rise and Decline of a Medical Philosophy*, Ithaca, Cornell University Press, 1973.

A. Vesalius, *Anatomicarum Gabrielis Falloppii Observationum Examen, apud Franciscum de Franciscis Senensem*, Venice, 1564.

A. Vesalius, *De humani corporis fabrica*, Basel, 1543.

A. Wear, *Galen in the Renaissance*, 1981.

E. Wiedemann, " Zu Ibn al Haitams Optik ", *Archiv für die Geschichte der Naturwissenschaften und der Technik*, 3, 1913, 1-53.

G. Whitteridge, *William Harvey. An anatomical disputation concerning The Movement of the Heart and Blood in living creatures*. Translated with introduction and notes, Oxford, Blackwell Scientific Publications, 1976.

FIGURES

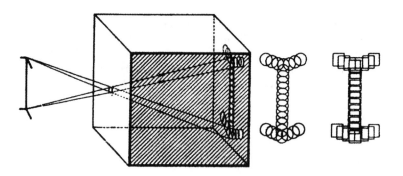

1. Camera obscur model

2. Kepler's Paralipomena

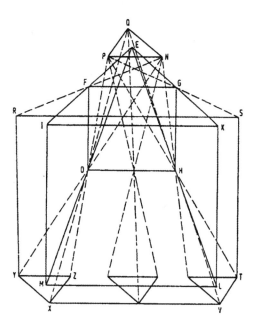

3. Kepler's camera obscura model

4. Dürer's perspective method

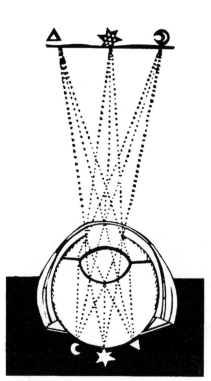

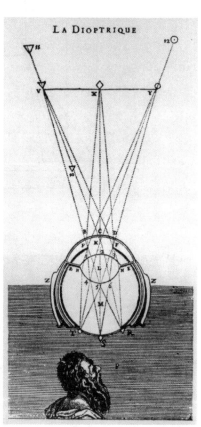

5. Kircher's eye model   6. Descartes' eye model

7. Frontespiece *De Visione*

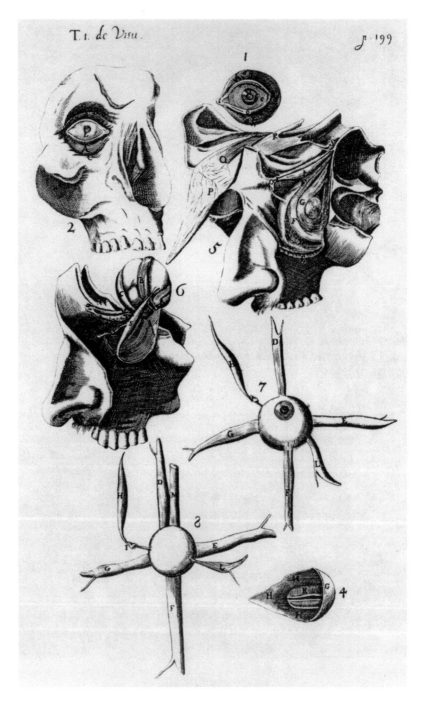

8. Frabricius'eye diagrams 1

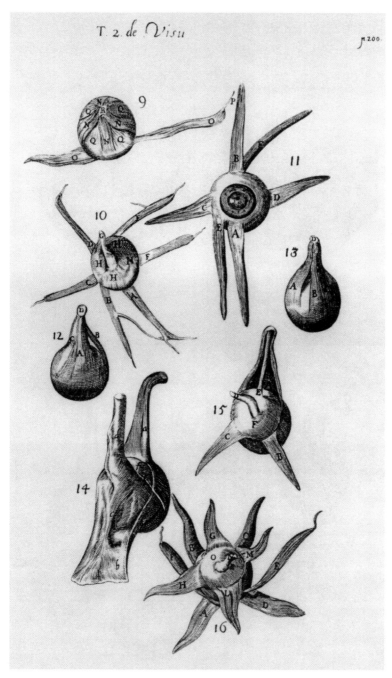

9. fabricius'eye diagrams 2

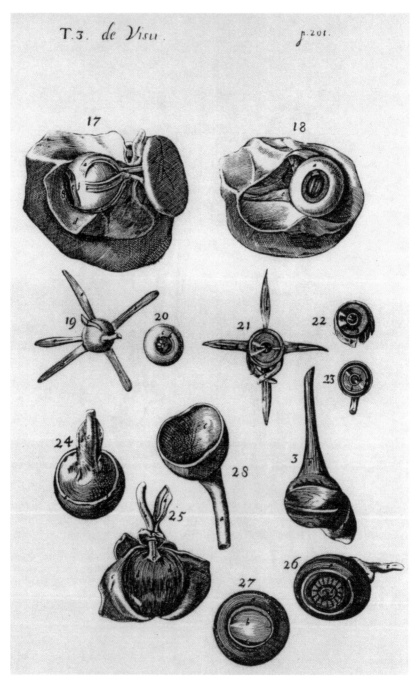

10. Fabricius'eye diagrams 3

11. Fabricius'eye diagrams 4

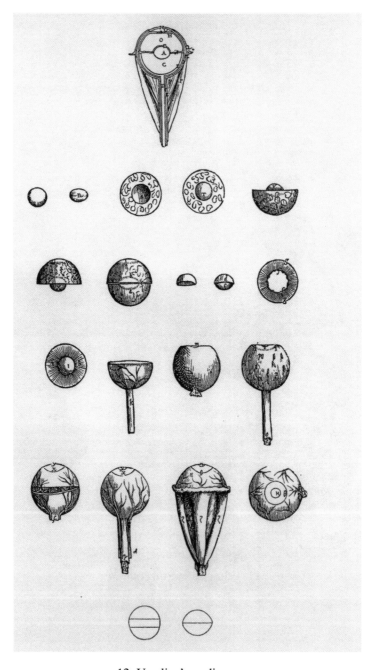

12. Vesalius'eye diagrams

13. Anatomical theatre of Padua

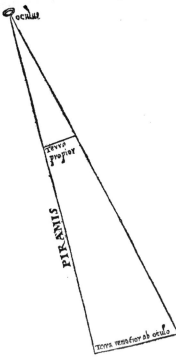

14. The visual pyramid

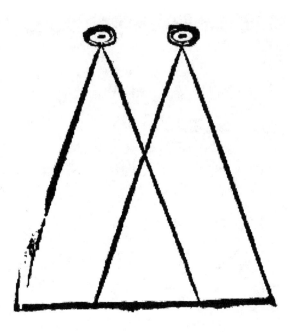

15. Biocular disparity

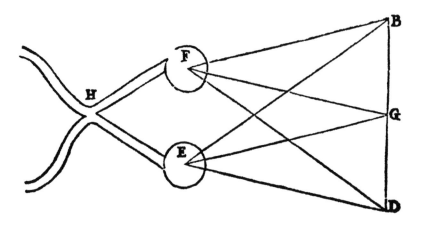

16. Danti's interpretation

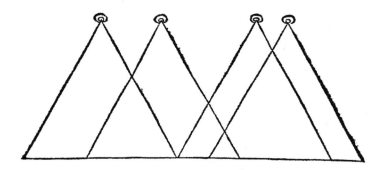

Oculi magis inuicem distantes.   Oculi magis inuicem propinqui.

17. The appropriate position of the eye

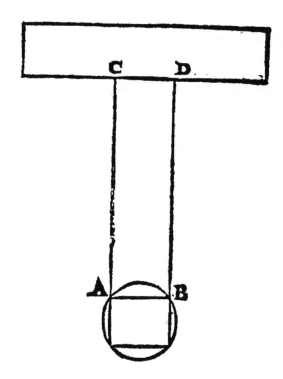

18. The spherical shape of the eye

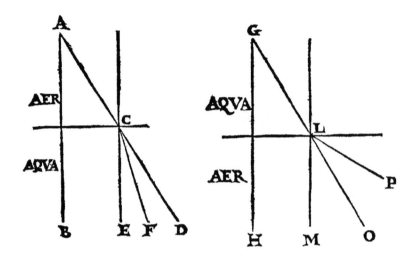

19 & 20. Diffractive media

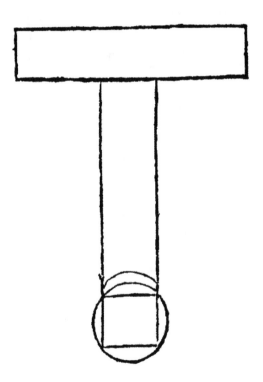

21. The spherical shape of the cornea

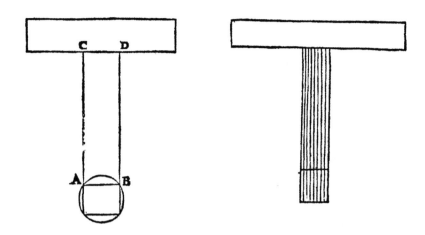

22 & 23. The inapt flat eye surface

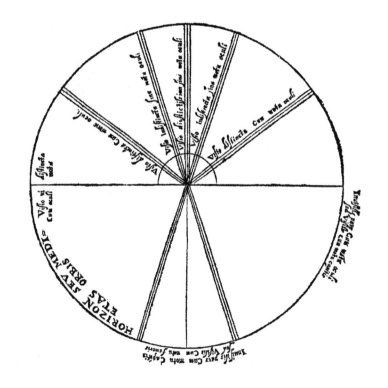

24. Variable range of visual field

HVMANI

1. Centrum Oculi.

2. Centrum Ara nea.

3. Centrum Cornea.

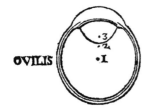

OVILIS

1. Centrum Oculi.

2. Centrum Aranea.

3. Centrum Cornet.

25. Eye of sheep and human being

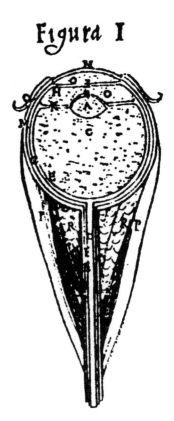

26. Valverde's eye diagram

berauit eos dns. Custodit dns omia ossa
eor:'unu ex hus n conteret. ws pecaror
pessima:'q' qui oderut iustu delinquent.
Redimet dns animas seruor suor:'e' n de
linquent onis qui sparit in eo.
Uldica dne nocentes me:'expugna inpug
nantes me. pprehende arma e' satum:'
e' exsurge in adiutoriu m. Effunde frame
am e' conclude adversus eos qui psequunt
me:'dic anime mee salus tua ego sum.
onfundant e' reuereant:'querentes anima
mea. Uertant retrorsu e' confundant:'co
gitantes m mala. Fiant tanquam puluis
ante faciem uenti:'e' angls dni coartans eos.
Fiat uia illor tenebre e' lubricu:'e' angls dni
psequens eos. Qm gratis absconderunt m
interitu laquei sui: superuacue exprobauerunt
anima mea. Ueniat illi laqueus que igno
rat:'e' captio qua abscondit apprehendat
eum:'e' in laqueum cadat in ipso ·

27. Ghent Psalter with bird dragon

28. Bird wearing spectacles (detail)

KEPLER'S CRITIQUE OF THE MEDIEVAL PERSPECTIVIST TRADITION

Raz Dov CHEN-MORRIS, Sabetai UNGURU

The historiographical problem of defining the originality of Kepler's theories of vision has reached a disturbing *status quo* during the last decade. On the one hand, it is admitted that there is novelty, or even a dimension of revolutionary breakthrough, in Kepler's ideas about vision. Yet, on the other hand, lines of continuity are clearly drawn from medieval systems and theories that lead to Kepler as their direct heir and descendant[1]. First, it has been established that Kepler's geometrical tools were not so much different from those of Alhazen, Witelo, or for that matter, Euclid. Furthermore, most of his ideas are to be found either directly in the medieval perspectivist tradition (e.g., a theoretical commitment to the rectilinearity of light's rays[2]) or in Neoplatonic speculations. Historiographical explanation is forced then to resort to Kepler's personal ingenuity, and to the fact that he was a better geometer than his predecessors (Alhazen, Witelo, Euclid).

---

1. The main contributions to this debates are : S.S. Straker, *Kepler's Optics : A Study in the Development of Seventeenth Century Natural Philosophy*, Ph.D. diss., Indiana University, 1970 ; A.C. Crombie, " The Mechanistic Hypothesis and the Scientific Study of Vision ", *Proceeding of the Royal Microscopical Society*, II (1967), 1-112. Lately, Crombie repeated his arguments, with little more to support them, in his " Expectation, Modelling and Assent in the History of Optics : Part I. Alhazen and the Medieval Tradition ", *Studies in History and Philosophy of Science*, vol. 21 n° 4 (1990) 605-632, and " Part II. Kepler and Descartes ", *Stud. Hist. Phil. Sci.*, vol. 22, n° 1 (1991), 89-115. D.C. Lindberg, *Theories of Vision From Al-Kindi to Kepler*, Chicago, University of Chicago Press, 1976 ; also his " Laying the Foundations of Geometrical Optics : Maurolico, Kepler, and the Medieval Tradition " in *The Discourse of Light from the Middle Ages to the Enlightenment*, William Andrews Clark Memorial Library, U.C.L.A., 1985 ; and his " The Genesis of Kepler's Theory of Light : Light Metaphysics from Plotinus to Kepler ", *Osiris*, 2, (1986), 5-42.

2. It is S. Straker contention that Kepler's main achievement in optical theory was the establishment of a commitment to the rectilinearity of light rays as a basis for the process of the " mathematization of nature ". However, as Lindberg showed in several places, except in specific problems, this commitment was already established by Euclid and was reinforced in the medieval perspectivist tradition.

It is our contention that this predicament of historiographical research is the result of an embedded tendency of the history of science to concentrate on a comparative study of disembodied ideas. The scientific tradition embedded in texts, leads as if " naturally " to an effacement of its particular moments in favour of the historian's beloved general continuities of " unproblematic " inter-textuality. This procedure is aided by the fact that scientific texts (esp. of early modern science) contain citations and point to influences meant to present their author both as the harbinger of novelties, and as the faithful follower of an ongoing scientific tradition.

In order to avoid this historiographical pitfall, we suggest a reconstruction of Kepler's particular point of view of the perspectivist tradition : How does Kepler present this tradition, and where does he use perspectivist ideas and arguments ? How does he rewrite and reconstruct his scientific tradition ? Where does Kepler assume a revolutionary position and where does he present himself as the apex of a tradition ? These questions assume that for Kepler the perspectivist tradition was not a completed whole, stagnant and a matter of the past. On the contrary it was something dynamic and a matter of Kepler's pointed interest.

Thus, we aim to put Kepler's intellectual endeavours in the context of his self-fashioning as a reader and writer. That is, we are concerned with the ways in which he faced a textual heritage, moulded it, and used it for his own projects. Kepler's new modes of reading of the perspectivist tradition enabled him to redraw the scientific boundaries between geometry and nature. Thus, he was able to apply geometry to physics by redefining geometrical objects as the representation of abstract quantities instead of direct abstractions of material bodies.

The medieval intellectual practice has concentrated on commentary and interpretation. The Renaissance added to that other modes of scientific reading and writing. One of these was the editing, and if needed emendation and reconstruction of ancient scientific texts. Paul L. Rose points out that Francesco Maurolico, for example, formulated his program as aiming towards " a restoration of certainty to mathematics ". This was to be accomplished initially by " restoring the good and true traditions, which for Maurolico are best represented in Euclid, Apollonius and Archimedes ". The difficulty Maurolico encountered in achieving that, consisted in the corruption of the texts of Greek mathematics. Their restoration was the *sine qua non* for the establishment of mathematical certainty, and thus an urgent and necessary *desideratum*[3].

A second Renaissance practice for assimilating authoritative texts was the imitation of an ancient scientific text ; this in accordance with the Ciceronian ideals which ascribed intellectual and stylistic authority to ancient textual pat-

---

3. P.L. Rose, " The Italian Renaissance of Mathematics : Studies on Humanists and Mathematicians from Petrarch to Galileo ", *Travaux d'Humanisme et Renaissance*, CXLV, (1975), 178.

terns and models. Structuring one's discourse after those patterns legitimized it *vis-à-vis* reality and truth. Copernicus's astronomical novelties did not merely reject the authority of Ptolemy. Copernicus presented himself as following ancient models and ideas, a move that legitimized the transformation of the Ptolemaic system[4].

In what follows we argue that Kepler presents a new mode of reading and a new way for the production of texts. The classical sources are no longer origins, nor are they the authoritative suppliers of textual patterns. They are historicized as instances in a tradition with no exceptional claim to truth.

Thus, Kepler has to introduce new grounds for his own claims for truth. This is why he must formulate anew the origins of the science of optics as the obvious foundation to astronomy. In so doing the classical texts become mere triggers, an opportunity for Kepler to initiate his own discourse. As the new optical discourse unfolds, it leaves the classical authorities behind and a new textual landscape comes in view.

This process appears most conspicuously in the way Kepler re-interprets the role of the perpendicular in his theory of reflected and refracted images. We shall concentrate on the way Kepler relates to the ancient and medieval authorities in his treatment of catoptrics. In these cases his intricate and ambiguous relationship to ancient textual authorities as sources of knowledge is fully revealed. On the one hand he rejects them as erroneous. Thus mere textual interpretation is discarded as a valid method in the mathematical sciences. Instead he attempts to establish *a priori* knowledge of optical truth. However, while this leads him to the verge of new discoveries, he is unable to make full use of his novel approach. Kepler is committed to that part of the classical tradition which he takes to be the phenomena of optical research. This prevents him from accepting the validity of a law of refraction based on complex proportions that do not result from the numerical values given by Witelo for the angles of refraction (the accepted phenomena) and which, furthermore, lack any visual (i.e., simple geometrical) equivalent.

The third chapter of Kepler's *Supplement to Witelo, or the Optical Part of Astronomy* deals with the foundations of catoptrics and the place of images. Its first part aims to refute what is wrong in the tradition of classical and medieval authorities. Kepler states " The catoptrical demonstrations of the opticians are still obscure in the foundation itself, while their derivation from the senses is itself to be demonstrated "[5]. Kepler emphasizes this point that " there is no error that is not derived from that origin "[6]. These statements reject from the

4. For Kepler's own estimation of Copernicus's indebtedness to Ptolemy *cf.* Kepler, *GW*, vol. 1, 50 and vol. 3, 141. For a modern reassessment of this estimation *cf.* D.J. Price, " Contra-Copernicus ", in M. Clagett (ed.), *Critical Problems in the History of Science*, Madison, Univ. of Wisconsin Press, 1969, 215.

5. Kepler, *GW*, vol. 2, 61.

6. *Ibid.*

outset Maurolico's program for the reform of classical mathematical knowledge. It is not enough to correct the texts and to supply the scientific community with better and reformed versions of the sources, since the sources in themselves are erroneous. This point is further explicated, as Kepler presents one of Euclid's errors in his *Catoptrics*, and shows that the error is in Euclid's axioms (i.e., the whole Euclidean system is to be rejected). Euclid asserts that whatever " falls under vision is seen according to its perpendicular to the surface of the mirror "[7]. Euclid proves this by relying on an axiom that states that if the point on the mirror where the perpendicular from the visual object hits the mirror is covered the object will not be seen, even though the rays from the eye to the point of reflection are unobscured. Kepler does not aim to correct this mistake and thus to save the classical foundations of *Catoptrics*. Instead, Kepler criticizes the Euclidean move thus showing that the whole optical tradition ever since Euclid is in error.

Kepler develops his argument and claims that Euclid has no other proof for the superiority of the perpendicular than the mistaken axiom. Euclid's only resort is to talk in an artificial fashion of the visible rays, and thus to ascribe to the perpendicular line some occult " skill " that enables the production of an image, when it hits the mirror's surface. However, Kepler contends that *veritas hoc more philosophandi vim patitur*. The outcome is that by assuming artificial species *ignorantia... tyrannidem stabilit*[8].

Alhazen and Witelo, claims Kepler, have understood the absurdity of the Euclidean axiom, since they omitted it from their treatises (although they follow Euclid in other respects). Witelo may have tried to supply the cause for the role of the perpendicular ray in producing an image in a mirror, but his failure made things worse since *obscuritas vero rei etiam hallucinationem illi attulit*[9]. Against Witelo's explanation that connects the place of the image with the place of the object, Kepler argues that one cannot deduce the place of the image only from the place of the object. A demand like that of Witelo for a complete correspondence between the object and its image would ruin the whole structure of *Catoptrics*, since there are " many differences between the

---

7. *Ibid.*, 62.

8. *Ibid.*

9. Kepler probably refers to Witelo's sentence " Furthermore, as is clear by definition, the form that appears in a mirror is the image of the object of sight. Therefore, that image must appear uniformly situated with respect to the point on the object of sight and the mirror, because otherwise the form would not be seen as an image ". This is related to the Witellonian assertion : " That it is only along the cathetus of incidence that a point on an object of sight maintains a uniform position with respect to every mirror from which its form is reflected. Furthermore, as is clear by definition, the form that appears is the image of the object of sight. Therefore, that image must appear uniformly situated with respect to the point on the object of sight and the mirror... So it will necessarily appear on the cathetus of incidence ". A.M. Smith, *Witelionis Perspectivae Liber Quintus : Book V of Witelo's Perspectiva* — An English Translation with Introduction and Commentary and Latin Edition of the First Catoptrical Book of Witelo's Perspectiva, *Studia Copernicana*, XXIII, Wroclaw *etc.*, 1983, prop. 36, 119-20.

object and its image ". In other words Kepler argues that since the place of the image does not necessarily follow exclusively from the place of the object, the latter has no power by itself to determine the existence of the image[10].

Whereas Witelo turns to Euclid's postulates for help (which Kepler has already rejected as false), Kepler tries to see whether Alhazen may save the argument. Alhazen attempts to explain from experience the fact that the image is seen always along the perpendicular falling from the visual object to the surface of the mirror. However this attempt fails and Alhazen ends up with an obscure sentence : " [to understand] the mode of natural things one is to look back to the position of their principles, and the principles of natural things are obscure "[11]. Kepler rightly comments that at the same moment when Alhazen is to give the causes for the phenomenon, he evades into saying the cause is occult and Kepler comments : *hoc non est demonstrare.*

A second possibility to understand Alhazen is to assume he falls back on some preconceived natural order of creation, or as Witelo attempted, on the order within the soul[12]. For Kepler these attempts fail because " all these visual affections are drawn from material necessity, where there is no place for consideration of final [causes] or of beauty "[13]. What does Kepler mean by material necessity ? In the next paragraphs Kepler explains his demand. The argument out of considerations of beauty assumes the perpendicular to have some special quality about it, some essential power to present the right size of the visual object. Therefore an argument aiming at beauty fails to explain *quid causae sit, cur potius in plano vera quantitatis habeatur, quam in curvo.* The main question Kepler poses to the perspectivist tradition relates to the way the geometrical shapes of the different mirrors corresponds to the different sizes of images. The question hints at where Kepler intends to find his answer. In contrast to the emphasis laid by Alhazen and Witelo on the power and strength of the perpendicular, Kepler asks for a reciprocal causal connection between the geometrical shape of the physical surface and the geometrical description of the passage of light.

10. Kepler, *Ibid.*

11. *Et cum oporteat ipsam comprehendi in linea reflexionis, ...comprehendetur in concursu huius lineae cum hac perpendiculari. Iam ergo assignauimus causam huius rei. Verum rerum naturalium status respecit situs suorum principiorum, & principia rerum naturalium sunt occulta.* Alhazen, " De aspectibus ", in *Opticae Alhazeni Arabis libri septem...* Edited by Friedrich Risner, Basel, 1572. Reprint, with an introduction by D.C. Lindberg, New York, 1972. Bk. V, prop. 10, 131. (my emphasis).

12. A.M. Smith, *Ibid.*, prop. 18, 100. Witelo contends that " Everything that is seen in any mirror is comprehended by sight along the shortest possible line ". Witelo relies on a former prop. V, 5 that nature acts along the shortest paths in everything she does and that " the reflection of forms from the surfaces of mirrors to the eye is completely natural, since we know (by the whole of Bk. IV) that it is produced by the action of nature and, like every other [type of] vision, is completed by the action of the soul. In fact, the soul corresponds to nature in animals ".

13. Kepler, *Ibid.*, 63.

The traditional relationship between the mathematical domain and physical reality was set by the Aristotelian formula that : " While geometry investigates natural lines but not *qua* natural, optics investigates mathematical lines, but *qua* natural, not *qua* mathematical "[14]. This formula was somewhat turned on its head in the medieval formula : " Every line along which light... reaches... is a natural sensible line... whithin which a mathematical line is to be assumed imaginarily "[15]. Witelo proceeds : *neque cadit lux minima in punctum mathematicum... sed in punctum sensibilem.* Aristotle emphasizes that the optician talks about mathematical lines as if they were concrete physical entities with physical effects. In contrast, Witelo brings the opticians' practice closer to that of the mathematicians who abstract from material forms geometrical forms. However the principle remained the same — the two domains are separated and the origin of geometrical shapes is by abstraction from concrete bodies located in the physical-material realm[16]. Kepler reads this move as an unnecessary separation between the mathematical aspect of reality and its material necessities and effects. In the first chapter of *A Supplement to Witelo* Kepler emphasizes that the lines considered by the opticians are not abstractions from material phenomena but representations of the motion of a two dimensional (i.e., mathematical) surface[17]. Thus the treatment of light under the laws of geometry is not an artificial discourse ideally imagined, but must include physical necessities and effects as well. Kepler does not assume a mechanical — materialistic science, where objects affect one another according to their material mass in movement. The basis of Keplerian science is in his declaration that material objects tend to assume geometrical shape, and these geometrical shapes are the main factor in Keplerian optics. Light is transmitted by all objects through their geometrical dimensions[18].

Thus, light as a geometrical surface interacts with physical objects not through their material qualities, but only through their geometrical surfaces :

---

14. Aristotle, " Physics ", Bk. 2, 194a 10-11, (trans.) R.P. Hardie and R.K. Gaye, in *The Complete Works of Aristotle*, J. Barnes (ed.), Princeton, New Jersey, Princeton University Press, 1984, vol. 1, 331.

15. *Omnis linea qua pervenit lux... est linea naturalis sensibilis... in qua est linea mathematica ymaginabiliter assumenda.* S. Unguru, *Witelionis Perspectivae liber secundus et liber tertius* : A Critical Latin Edition with an English Translation, Introduction, Notes and Commentary, *Studia Copernicana*, XXVIII, Wroclaw *etc.*, 1991. Bk. II prop. 3, 46, 240.

16. That this inverted understanding of the Aristotelian formula was still accepted in the early 17th century is seen from the opening lines of Scheiner's preface to his *Oculus, hoc est, fundamentum opticum*, Innsbruck, 1619, *Physici quam Optici circa visibilia, & organum visus versantur ; modo tamen diverso. Geometria enim, teste Philosopho, l. 2 Phys. t. 20 de Physica linea considerat, sed non quatenus est Physici : Perspectiva autem mathematicam quidem lineam, sed non quatenus Physica est.* See also P. Dear, *Discipline and Experience, The Mathematical Way in the Scientific Revolution*, Chicago, London, 1995, 51-57.

17. *Nam radius... nihil aliud est nisi ipse motus lucis... in luce motus ipse est recta... linea, mobile vero, est superficies quaedam, Ibid.*, 21.

18. *Dictum enim est, debuisse illam communicari corporibus omnibus. ea communicatio fieri debuit dimensionum coniunctione, Ibid.*, 20.

" Light is not impeded by the solidity of the bodies, inasmuch as they are sol-ids, because it can least pass through them... Light is affected by the surfaces of any opposed bodies "[19]. Summarizing this point Kepler argues that the def-inition of light has two aspects : one relates to what is the essence of light, the other relates to its quantity. The first is the ability to illuminate and the other is local motion. The motion of light (i.e., the expression of the quantitative aspect) has two objects : the medium, and the physical object that it meets. These two objects affect light by refracting it (the media) or by reflecting it (the polished surface). However, these effects concern the motion of light, that is they are quantitative effects and are created by the interaction of geometrical surfaces.

It is Kepler's challenge to prove according to these principles basic optical theorems, such as that the angle of reflection is equal to the angle of incidence. Kepler wants to show that what affects the motion of light is not any occult quality of light, or of " Nature " in general, but the geometrical shapes assumed by material bodies. Kepler's proof is divided in two steps. First he establishes that light flowing to a surface is reflected to the opposite side. Then, he dem-onstrates that the angle of incidence is equal to the angle of reflection. In both cases Kepler uses elements which were applied already by Alhazen and Witelo, that is the analogy to a solid body that hits a solid surface, and the composition of oblique motion from perpendicular and parallel surfaces[20].

However, Kepler re-interprets these elements. For Alhazen the comparison of the reflection of light to the rebounding arrow from a solid surface is limited and serves as a simile not an identity. In the words of Lindberg : " these com-parisons [were]... meant to elucidate the geometry and causes of reflection and refraction but not the nature of the reflected or refracted entity "[21].

Kepler, on the other hand, isolates the element of motion and states the case as a general rule of anything that strikes violently an opposing object[22]. The second element (i.e., the decomposition of oblique motion) serves Alhazen and Witelo as an artificial and heuristic construct, the true cause for reflection according to equal angles is that Nature tends to operate according to the short-

19. *Ibid.*, propositions 10 & 12, 22.

20. For a discussion of these elements in the pre-Cartesian perspectivist tradition see : A.I. Sabra, *Theories of Light From Descartes to Newton,* London, 1967, 71-78. However, Sabra overlooks that Alhazen and Witelo, while discussing the analogy to a rebounded ball and the decomposition of oblique motion, still give the reason from " minimum path " for the equality of angles of incidence and reflection, see the following notes 26, 27.

21. D.C. Lindberg, *Theories of Vision : From Al Kindi to Kepler,* 80.

22. *Cur autem et in motu physico et in luce accidat repercussus, causa est in motus violentia. Cum ergo vis movens non omnis a conflictu aboleri potest ; superabit itaque motus terminum suae lineae, superficiem scilicet. At non potest in directum : obsistit enim illic corpus corpori : hic superficies superficiei : illic in solidum, hic ex parte, ut audiemus. Relinquitur igitur, ut in opposi-tum,* Kepler, *Ibid.,* 25.

est path[23]. Kepler rejects this line of reasoning. The geometrical model is constructed as a representation of possible paths for the motion of light, and their interaction with other physical surfaces creates the true and real linear path of the motion of light[24].

Kepler applies the same analysis and separation of the element of motion from the other qualities of light also in the case of refraction. " For as in the case of physical motion a spear is sometimes made to collide with a thing towards which we aim it and, adhering together, they both proceed on the same path with one motion. So the same thing happens in the case of light and the more dense surface which light penetrates, yet without material or the dimension of solidity "[25]. Towards the end of the same proposition Kepler clarifies this point even more emphatically : " This comes into use at all inclinations... And thus this effect of physical violent motion likewise is copious in [the case of] light with its own kind "[26]. Furthermore, Kepler's analysis of refraction compares the motion of light to oars in a boat and to theories of the balance. Thus he can separate the element of motion from the mobile and turn geometrical lines into representations of possible paths within a geometrical model (either a circle in case of a balance, or a table in case of the oars). The perpendicular drawn within that model is a limiting case : in the case of the balance the perpendicular is drawn to the centre of the balance and to the centre of the circle described by it. This perpendicular is the limit for the ascension or descension of the balance ; in the case of the oars it is that line to which the oars deflect in case of an inclined violent motion of the river. In the case of light the density of the medium limits the dispersion of light. The oblique ray cannot proceed in the same direction, nor can it disperse in a direction further away from a perpendicular ray falling from the same source). Thus, an oblique ray must be refracted from its point of incidence somewhere between the path of a ray passing the denser medium perpendicularly and the continuation of its original oblique path (i.e. it will be refracted towards the perpendicular).

23. Cf. Witelo, Ibid., Bk. v, prop. 5, and prop. 18. Witelo, following the traditional scheme ascribes the true physical cause for the equality of the angles in reflection to the principle of " the minimum path ". Thus the geometrical composition of parallel and perpendicular rays has a secondary status vis-à-vis causal account : " Now we know (by v, 5) that nature acts along the shortest paths in everything she does... Everything that is seen in any mirror is comprehended by sight along the shortest possible line ". (A.M. Smith, Ibid., 101).

24. Our differentiation between " abstraction " and " representation " has some affinity with Foucault's description of the move from the Renaissance episteme based on similitude to the 17th century Classical episteme based on representation. Whereas ancient and medieval mathematical optics assumed that the lines are similar to real objects (i.e., they are like true physical shapes but not exactly the same), Kepler's mathematics posits geometrical entities to be the true representations of possible paths in the world. Cf. M. Foucault, The Order of Things : An Archaeology of the Human Sciences, New York, 1970.

25. Kepler, Ibid., 27-28. (trans. S.M. Straker, " Kepler's Optics : A Study in the Development of Seventeenth Century Natural Philosophy ", Ph.D. diss., Indiana University, 1970, vol. 2, 535, my emphasis).

26. Kepler, Ibid., 31.

All physical motions can be represented by geometrical lines. The peculiar aspect of light is that its motion can be described with no reference to matter, since light is an incorporeal being. The tendency of the refracted ray of light towards the perpendicular is not because of an innate virtue of light or of perpendicular lines, nor because of any diminution in the quantity of light in the oblique ray. The only part played by the perpendicular is its being a limiting case of direct radiation[27] : i.e., the perpendicular line must be taken into account as a representation of one of the possible paths for a ray of light that falls to the surface of a denser (or rarer) medium. The perpendicular is neither an abstraction of some concrete entity, nor a sign, or a conventional symbol, of something that was there before (the way written letters are signs for the sound once uttered), nor an ideal entity existing in some non-sensual realm. The perpendicular (or any line for that matter) is a sort of representation. The line is described and created by the possible movements of any body in the world (including the two dimensional surface of light). Thus, Kepler's conception of geometrical physics was more than mere Renaissance Platonism. Kepler's geometry did not constitute a separate realm, but was created by the actions and passions of the physical world itself. Furthermore, while most of the Keplerian moves are not original with him, but can be found in his medieval and classical sources, he does not only re-interpret these moves, but purges the language of perspective from any recourse to occult and inherent qualities of light and geometrical lines. The natural world does not aspire to imitate and fulfil imperfectly some ideal divine plan. On the contrary despite its blind and material motions, the natural world represents somehow directly and exactly the divine archetypes to the human eyes and intellect.

In the third chapter of *Ad Vitellionem*, Kepler applies this conception of geometry to explain the role of the perpendicular in determining the position of the image in case of reflection. The classical and medieval traditions had to fall back on some inherent virtue of the perpendicular in order to explain why the image in the mirror lies along the perpendicular from the visible point to the mirror's surface. Kepler's point of departure is the eye's visual distortion. The eye imagines the object to be where its refracted or reflected image is. Then, the eye imagines its line of sight to be in straight line to that image, and does not perceive the point of its refraction or reflection. The place of the image then is that point where the produced rays of vision from each of the eyes meet after they passed through the point of refraction or reflection. Each eye and its line of sight is on a surface of reflection (or refraction). The place of the image is where these two surfaces meet, and one point along their common section is the visible point. Now these surfaces are set perpendicularly

---

27. In another proposition Kepler states that *Atqui haec opera non sunt formae consilio utentis aut finem respicientis, sed materiae suis Geometricis necessitatibus astrictae, Ibid.*, ch. 3, prop. XVI, 71.

above the mirror, therefore their common section is a perpendicular line and one of its points is the visible object. The object and its image are on that common section between the two surfaces of the eyes, and this common section is the perpendicular from the visible point to the surface of the mirror. Kepler can claim now that he managed *Non igitur occulta natura lucis, non naturae universalis ingenium, sed sola visus latitudo, inter causas concurrit, cur visus imaginem reponat in perpendiculari*[28].

Thus the cause for the role of the perpendicular in determining the place of the images in a mirror is not due to some inherent characteristic, but because it represents the common geometrical section of the two surfaces on which the imaginary visual ray from each eye meet the surface of reflection (or refraction). Thus, the laws of geometry win the day, the boundaries between natural causes and geometrical ones having been transgressed. Instead of the classical and medieval landscape of concrete physical objects and their direct geometrical abstractions, Kepler presents a geometrical landscape produced out of possible lines and shapes that materialize by the real motions of physical bodies. Into this model created according to geometrical necessity, Kepler moulds ancient experimental results. Thus in his treatment of the conic sections the curves are no longer separated geometrical entities, each an abstraction of a specific section of the cone, but are parts of a systematic model whereby each receives its definition from its relations to the other curves. This geometrical model is to accommodate the different angles of refraction supplied by Witelo. Although this failed, Kepler still believed that this geometrical model is to give sense to experimental results. Thus, Kepler puts traditional authorities in a new position. Alhazen and Witelo are no longer models to imitate or to comment upon, but are the suppliers of experimental phenomena. Kepler does not argue with these phenomena, but with Alhazen's and Witelo's interpretation of them. He points to a new context of observational astronomy to show the inability of medieval optical theories to supply a satisfying account. To the ancient texts, that cannot supply anymore points of departure for Kepler's own writing, he adds his new conception of geometrical representational language of reality. Instead of commenting or imitating ancient tradition Kepler turns to a new kind of text composed of observational and quantitative results. The Keplerian Book of Nature is not a signifier of other divine texts, but the only true representation of the geometrical model conceived in the divine mind before creation. Thus, for instance chapter IV of *Ad Vitellionem* does not depart from a textual debate over interpretation but with a debate on observational data (i.e., the skirmish between Rothmanns and Tycho over atmospherical refraction[29]). The solution of this debate is to be found not in better texts but in a better geometrical model.

---

28. *Ibid.*, 73.
29. *De velitatione Tychonem inter et Rothmannum, super Refractionum negotio*, 77.

HOW GALILEO'S MIND GUIDED HIS EYE WHEN HE FIRST LOOKED AT THE MOON THROUGH A TELESCOPE

William R. SHEA

THE MOON'S OLD FACE

The nature of the Moon was a moot question for the Ancients and a lively account of the various theories that had been adduced was to be found in Plutarch's dialogue *The Face on the Moon*[1], written at the end of the first century. These theories include seeing the Moon as : (a) an indestructible and

---

1. Plutarch, *The Face on the Moon*, 930 D, in Plutarch's *Moralia* transl. by H. Cherniss and W.C. Helmbold (Loeb Classical Library), London, William Heinemann, 1984, vol. XII, 111. According to Antonio Favaro, Galileo owned a Latin version of the *Moralia* transl. by W. Xylander and published in Venice by Girolamo Scoto in 1572 (A. Favaro, " La libreria di Galileo ", *Bullettino di Bibliografia e di Storia delle Scienze Matematiche e Fisiche ... da B. Boncompagni*, XIX, n° 85 (1886), 245). Galileo could also have used an Italian version, *La faccia che si vede nella luna*, in *Opuscoli morali di Plutarco Cheronese Filosofo e Historico Notabilissimo... tradotti in volgare dal Sign. Marc'Antonio Gandino & da altri letterati*, Venice, Fioravante Prati, 1598. See P. Casini, " Il *Dialogo* di Galileo e la luna di Plutarco ", in P. Galluzzi (ed.), *Novità Celeste e Crisi del Sapere*, Florence, Istituto e Museo di Storia della Scienza, 1983, 57-62. Plutarch's work was considered important enough to be translated from Greek into Latin by Johann Kepler who added a substantial set of notes that are almost as long as Plutarch's text (*Plutarchi Philosophi Chaeronensis Libellus de facie, quae in orbe Lunae apparet*, and *Notae*, in J. Kepler, *Gesammelte Werke*, vol. XI, part 2 (ed. by V. Bialas and H. Grössing), 380-409, 410-436. Kepler's translation was published posthumously as an appendix to his *Somnium* in 1634. As early as 1604, Kepler quoted Plutarch several times in his *Ad Vitellionem Paralipomena quibus Astronomiae Pars Optica Traditur* (in J. Kepler, *Gesammelte Werke*, ed. by M. Caspar, F. Hammer et alii. Munich, C.H. Beck'sche Verlagsbuchhandlung, 1938-1993, vol. II, 201 ff.). When he saw Galileo's *Sidereus Nuncius*, his immediate reaction was to tell the Tuscan ambassador in Prague, Giuliano de' Medici, that Galileo " had defended Plutarch with very persuasive arguments " (letter of Giuliano De' Medici to Galileo, 19 April 1610, in Galileo Galilei, *Opere*, ed. by A. Favaro, 20 vols. Florence, G. Barbèra, 1899-1909, vol. X, 348). He shortly thereafter informed Galileo that he had read Plutarch as early as 1593 on the advice of Michael Maestlin, his professor of astronomy at Tubingen, and that he had begun to translate *On the Face of the Moon* before hearing of Galileo's celestial discoveries (Kepler, *Dissertatio cum Nuncio Sidereo in Opere di Galileo Galilei*, vol. III, part 1, 112). The Italian mathematician, Camillo Gloriosi, also thought of Plutarch when he read Galileo's work. " What Galileo says about the Moon ", he wrote to a friend, " is nothing new. It was already stated by Pythagoras, and Plutarch wrote a book on the topic " (letter to Giovanni Terrenzio, 29 May 1610, in Galileo, *Opere*, vol. X, 363).

unchangeable sphere made of a substance unknown on Earth (the position of Aristotle) ; (b) a smooth, translucent crystalline sphere ; (c) a body of condensed fire ; (d) a mirror reflecting the terrestrial ocean and continents ; and (e) a physical body like the Earth with mountains, valleys and seas. Arguments for one or another view involved such questions as whether celestial matter was different from earthly matter or how the Moon could shine. One of the speakers, Lucius, states that " the moon is very uneven and rugged with the result that the rays... (are) coming to us, as it were, from many mirrors ", and later the narrator (presumably Plutarch himself) notes that, during an eclipse, changes of colour are visible along the length of the advancing shadow. This plainly indicates " that the Moon has not a single plane surface like the sea but closely resembles in constitution the Earth " and, hence, has valleys and mountains[2].

In one passage the opinion that the Moon is similar to the Earth is said to " turn things upside down ", just like the theory of Aristarchus of Samos who was accused of impiety for " disturbing the hearth of the universe by trying to save the phenomena on the assumption that the heaven is at rest, and that the Earth revolves along the ecliptic and at the same time rotates about its own axis "[3].

The association of ideas between the heliocentric theory and the terrestrial nature of the Moon in Plutarch probably encouraged Galileo to examine possible connections. When he was appointed professor of mathematics at the University of Padua in 1592, Copernicanism was almost half a century old. Discussed by professional astronomers, and occasionally derided by theologians like Luther, it had yet to fire the imagination of the general public. If anything, it brought a smile to the lips of the common man who stood with his two feet solidly planted on the firm earth. Why if the Earth were spinning like a top, everything on its surface would fly off ! Public buildings, churches, hospitals, schools, as well as astronomical observatories would be swept away ! The clouds and the Moon would be left behind. The professors, who heard these things, smiled back and said deep things about sympathy, affinities and attraction between the Earth and the Moon. The wiser ones remained silent and pinned their hopes on the heavens. Their prayers were answered by the appearance of a nova in 1604. The new star caused a considerable stir among students

---

2. *The Face on the Moon*, 934 F, *Ibid.*, 139-141.

3. *The Face on the Moon*, 923 A, *Ibid.*, 55. *Sidereus Nuncius*, Drake trans., 23 (In Galileo, *Opere*, vol. III, 62, Van Helden trans., 40). Although Plutarch clearly spoke of dark and light regions of the Moon, no image of the Moon itself, as it presents itself to the eye, existed in Western culture as late as 1400. The first naturalistic drawing of the Moon appears in the *Crucifixion* painted around 1420 by the Flemish master, Jan Van Eyck, and now in the Metropolitan Museum in New York, as was shown in 1994 by S.L. Montgomery, " The First Naturalistic Drawing of the Moon ", *Journal for the History of Astronomy*, XXV (1994), 317-320. Until then, scholars had attributed the first naked-eye drawings to Leonardo da Vinci, specifically to three sketches that appear in his notebook (G. Reaves and C. Pedretti, " Leonardo da Vinci's Drawings of the Surface Features of the Moon ", *Journal for the History of Astronomy* , 18 (1987), 55-58.

in Padua, and Galileo gave three public lectures to large audiences in which he explained that the absence of any apparent displacement of the new star against the background of fixed stars (what is technically called parallax) indicated that the new star had been generated beyond the lunar region, namely in that part of the cosmos that the Aristotelians held to be immune from change.

Matters might have rested at this level of general conjecture had not something new occurred. This time the novelty did not descend from the ethereal regions of speculation ; it was rather the mundane but felicitous outcome of playing around with lenses, in Italy around 1590, in the Netherlands in 1604, and in the whole of Europe by the Summer of 1609[4]. Out of a toy to make objects appear larger, Galileo made, first, a naval, and then a scientific instrument. He combined a concave and a convex lens, and the result was the instrument still commonly used at the opera, and in theatres and concert halls, and known as an opera glass. Roughly at the same time, Kepler had worked out the focal distances of two convex lenses and had fitted them to enlarge objects. But the resulting image was upside down whereas Galileo's image was the right way up. This was no mean success, since seeing buildings standing on their head does not, at first sight, seem very promising even if they are enlarged.

Rumours of the invention of the telescope probably reached Galileo in July 1609 when he visited influential friends in Venice to explore ways of increasing a professional salary that had become inadequate to the needs of an elder brother expected to provide dowries for two sisters. He received little encouragement from the Venetian patricians who controlled the finances of the University of Padua, but he had a flash of insight when he heard that someone had presented Count Maurice of Nassau with a spyglass by means of which distant objects could be brought closer. The Venetians might not see how they could increase his salary, but what if he succeeded in enhancing their vision ?

When Galileo returned to Padua on 3 August his fertile mind was teeming with possibilities. By 21 August he was back in Venice with a telescope capable of magnifying eight times. He convinced worthy Senators to climb with him to the top of high towers from whence they were able to see boats coming to port a good two hours before they could be spotted by the naked eye. The strategic advantages of the new instrument were not lost on a maritime power, and it suddenly became clear to all that Galileo's salary should be increased from 520 to 1.000 florins per year.

Unfortunately, after the first flush of enthusiasm, the Senators heard the sobering news that the telescope was already widespread throughout Europe, and when the official document was drawn up it stipulated that Galileo would only get his raise at the expiration of his existing contract a year later, and that he would be barred, for life, from the possibility of subsequent increases. This

---

4. See A. Van Helden, *The Invention of the Telescope*, Philadelphia, American Philosophical Society, 1977 (*Transactions of the American Philosophical Society*, volume 67, part 4).

incident understandably made Galileo sour. He had not claimed to be the inventor of the telescope, and if the Senators had compared his instrument with those made by others they would have found that his was far superior. Let the Venetian Republic keep the eight-power telescope ! He would make a better one and offer it to a more enlightened patron ! Better still, he would show that much more could be revealed, not only on land and sea, but beyond the reaches of human navigation. It was natural that he should seek confirmation with the new fifteen-power telescope that he had assembled by November 1609. The fact that the mountains on the Moon cannot be seen with the naked eye is obvious to everyone, but equally clear to the astronomer, although sometimes forgotten by the historian of ideas, is the no less important fact that these mountains are also invisible to the eye that looks at them through a telescope. Their presence can only be inferred from variations in the light and dark areas of the lunar surface interpreted by someone with a knowledge of perspective. Most astronomers in Galileo's day were trained as mathematicians but he had also received instruction in the theory and practice of perspective, and he had distinguished himself as an amateur painter. According to his first biographer, Vincenzo Viviani, Galileo used to say to friends that had he been able when young to choose a profession, he certainly would have chosen painting. Indeed his talent for painting was so natural, and in time he acquired such excellent taste, that his opinion on paintings and drawings was preferred by professional painters to that of master painters. Cigoli, Bronzino, Passignano, Empoli, and other famous painters of his time, who were very good friends of his, asked his opinion. Recognizing in Galileo such perfect taste and supernatural grace in such a noble art, they asked him questions about the order of stories, perspective, color, and any other topic concerning painting that nobody else — not even professionals — was able to answer. The most renowned painter Cigoli, whom Galileo considered the greatest painter of his time, attributed a great deal of his achievement to Galileo's excellent teaching, and he was particularly proud of saying that in perspective Galileo had been his only teacher[5].

The perception of depth is linked to previous experience of chiaroscuro and the illumination of objects by a light source and slowly rotated. Galileo was fully apprised of this as we can see from what he later wrote to Cigoli : " We know of depth, not as a visual experience *per se* and absolutely, but only by accident and in relation to light and darkness "[6]. He argues that three-dimensional objects, such as statues, are seen in relief when illuminated by a light

5. V. Viviani, *Racconto Istorico della Vita di Galileo Galilei*, in Galileo, *Opere*, vol. XIX, 602. I quote from the translation by Michael Segre cited in I.B. Cohen, " What Galileo saw : the Experience of Looking Through a Telescope ", Proceedings of the Conference, " Galileo a Padova 1592-1610 ", *Occasioni Galileiane*. Trieste, Edizioni LINT, 1995, vol. V, 173.

6. Letter of Galileo to Cigoli, 26 June 1612, in Galileo, *Opere*, vol. XI, 341. The letter is discussed by E. Panofsky, *Galileo as Critic of the Arts*. The Hague, Martinus Nijhoff, 1954, 5-11. Panofsky recalls that Cigoli in his youth had been instructed in perspective and mathematics by the same Ostilio Ricci who taught Galileo (5, n. 2).

source because some parts appear bright and others, dark. It was because Galileo was familiar with the way an uneven surface is illuminated and casts shadows that he was able to recognise features of the lunar surface that had escaped others. But if he had not been looking for earth-like he would not have seen them[7].

THE MOON'S NEW LOOK

Galileo studied the Moon through his new instrument from 30 November 1609 until 19 January 1610, and he reported on his findings in the *Sidereus Nuncius* that came off the presses on 12 March 1610 and was immediately acclaimed. What struck him was a large number of spots that were smaller than the " large or ancient ones " that are visible to the naked eye. These smaller spots, he added with justifiable pride, " had never been seen by anyone before me ". This was the raw data of sense experience that he transformed, by analogy with what we see on Earth, into mountains and craters : " From observations of these spots repeated many times I have been led to the opinion and conviction that the surface of the Moon is not smooth, uniform and precisely spherical as a great number of philosophers believe it (and the other heavenly bodies) to be, but is uneven, rough, and full of cavities and prominences, being not unlike the face of the earth, relieved by chains of mountains and deep valleys "[8].

The similarity between the lunar surface and the face of the Earth may seem a foregone conclusion to the modern reader, but it went against a coherent and deeply-entrenched cosmology that considered the Moon as immune to the changes that we experience on Earth. The greatest Jesuit astronomer of the age, Christoph Clavius, came to recognise that Galileo had discovered real satellites orbiting around Jupiter, but when asked by Cardinal Robert Bellarmino to state his professional opinion about the unevenness of the lunar surface, he replied

7. S.Y. Edgerton, *The Heritage of Giotto's Geometry*, Ithaca, London, Cornell University Press, 1991, 223-234. Edgerton provides illustrations from Daniel Barbaro's *Pratica della Perspettiva* (1568) and L. Sirigatti's *Pratica di Prospettiva* (1596). See also M. Kemp, *The Science of Art*, New Haven, London, Yale University Press, 1990, 53-98, and E.A. Whitaker, " Selenography in the Seventeenth Century " in R. Taton and C. Wilson (eds), *Planetary Astronomy from the Renaissance to the Rise of Astrophysics*. Part A : *Tycho Brake to Newton*. Cambridge, Cambridge University Press, 1989, 122. For a general account of Galileo's telescopic observation, see G. Righini, *Contributo alla interpretazione scientifica dell'opera astronomica di Galileo*, Florence, Istituto e Museo di Storia della Scienza, 1978. Galileo's *Sidereal Messenger* is currently available in two English versions, one by S. Drake, *Telescopes, Tides and Tactics* : *A Galilean Dialogue about the " Starry Messenger " and Systems of the World*, Chicago, Chicago University Press, 1983, which is based on the partial translation that Drake had published earlier in *Discoveries and Opinions of Galileo*, Garden City, Anchor Book, 1957 ; the other by A. Van Helden, *Sidereus Nuncius or The Sidereal Messenger*, Chicago, Chicago University Press, 1989. There are two excellent French translations with introduction and notes, one by F. Hallyn, *Le messager des étoiles*, Paris, Seuil, 1992, the other by I. Pantin, *Le messager céleste*, Paris, Les Belles Lettres, 1992.

8. *Sidereus Nuncius*, Drake trans., 23-24 (In Galileo, *Opere*, vol. III, part I, 62-63 ; Van Helden trans., 40).

that he was unconvinced[9]. Discovering heavenly bodies that had always been in the heavens above, albeit unseen, was one thing ; claiming that the Moon was made of matter subject to alteration and decay was quite another. Clavius believed that the light and dark spots represented different condensations of the primal lunar matter, a view that had been discussed by Averroes in the Middle Ages. Galileo knew that he had to argue his case, and his strategy was to show that the phenomena observed can only be satisfactorily explained if the lunar surface is similar to the terrestrial one. Here are some instances.

First, if we look at the Moon when it is four or five days old (see Figure 1), we notice that the line that divides the shaded part from the lighted is not a smooth semicircle, as would happen on a perfectly spherical solid, but a rough and wavy line. Some lighted areas extend beyond the boundary into the dark-ened region, while some shaded areas or spots reach into the illuminated part. Galileo asked his reader to " note particularly that the said small spots agree always in having their darker parts directed toward the sun, while on the other side, opposite to the sun, they are crowned with bright contours, like shining summits ", and he brought the following analogy into play : " A similar sight occurs on earth about sunrise, when we behold the valleys not yet flooded with light although the mountains surrounding them are already ablaze with glow-ing splendor on the side opposite the sun. And just as the shadows in hollows on earth diminish as the sun rises higher, so those spots on the moon lose their blackness as the region illuminated grows larger and larger "[10].

Second, bright points appear beyond the boundary and inside the darker portion of the Moon. They are seen to increase in size and brightness, and after an hour or two become joined to the illuminated part that has meanwhile grown in size. Which prompts the rhetorical question, " On Earth before sun-rise are not the highest mountain peaks illuminated by the sun's rays while plains remain in shadow ? And when the sun is fully risen, does not the illu-mination of plains and peaks become finally united ? "[11].

Third, in the " ancient " large lunar spots that are even and uniform, brighter patches only crop up here and there. This might be taken as an objection against the theory that the moon is covered with cavities and prominences, but Galileo turns it into yet another argument in favour of a rugged lunar surface : " Hence, if anyone should wish to revive the old Pythagorean opinion that the Moon is like another earth, its brighter parts might very fitly represent the land

9. See the letter of the mathematicians of the Roman College to Cardinal Bellarmino, 24 April 1611, in Galileo, *Opere*, vol. XI, 93. The other Jesuit co-signatories of the letter, Christoph Grien-berger, Odo Malcotio and Giovanni Paolo Lembo tended to side with Galileo. Copies of this letter were circulated and one reached Lodovico delle Colombe who wrote from Florence to say that he shared Clavius' reservations (letter to Clavius, 27 May 1611, in Galileo, *Opere*, vol. XI, 118).

10. *Sidereus Nuncius*, Drake trans., 25 (In Galileo, *Opere*, vol. III, part I, 63 ; Van Helden trans., 40-41).

11. *Sidereus Nuncius*, Drake trans., 26 (In Galileo, *Opere*, vol. III, part I, 64 ; Van Helden trans., 41).

surface, and its darker regions that of the water. I have never doubted that if our globe were seen from afar, flooded with sunlight, the land areas would appear brighter and the watery regions darker "[12].

The statement that land patches would appear brighter than the seas caused surprise, and in the First Day of the *Dialogue on the Two Chief World Systems* that he published in 1632, Galileo has his spokesman, Salviati, pour water on bricks to show that this makes them look darker. When someone suggests that the smoother a surface, the more it is capable of reflecting light, Salviati hangs a mirror on a wall that is struck by the Sun, and shows that for most positions it appears darker than the wall itself[13].

Fourth, having reached this stage in the confirmation of his hypothesis, Galileo began to see evidence of mountains that went beyond the telescopic image. He literally perceived what his mind told him to see, as is revealed in the diagram at the top of Figure 2, where we notice in the upper half of the Moon a ridge of mountains in the shape of a hemicycle to the left of the boundary line. This contour in black invading the light portion is exaggerated and agrees poorly with the white contour in the second diagram which shows the Moon at last quarter, a day or two later.

Fifth, Galileo's drawings of the Moon at first and last quarter (see Figure 3) show a libration in latitude (an apparent oscillation whereby each of the lunar poles tilts in our direction in turn) of 9° vertically measured from a crater (later called Albategnius) that Galileo saw as a proof that there are mountains on the Moon. A comparison of the Moon at last quarter as seen through a modern telescope and as sketched by Galileo reveals that the size of the crater is greatly enlarged in Galileo's drawing (see Figure 4). Galileo noticed the difference between the illumination of the crater at first and last quarter, and he was anxious that the analogy with a terrestrial valley surrounded by hills should not be lost on his readers. Like many teachers, before and after him, he exaggerated the size of what he had observed in order to bring out the salient features. This was all the more necessary since, in a small woodcut, Galileo could not highlight the shifting pattern of shadows without giving the crater considerable width. He even invented a terrestrial counterpart to strengthen his analogy : the " perfectly circular cavity " [the crater Albategnius], he writes, " offers the same appearance as would a region like Bohemia if that were enclosed on all sides by very lofty mountains disposed exactly in a circle "[14]. Here it is the Earth that is rendered Moon-like for illustrative purposes !

12. *Sidereus Nuncius*, Drake trans., 27 (In Galileo, *Opere*, vol. III, part 1, 65 ; Van Helden trans., 43).

13. Galileo Galilei, *Dialogo sopra i due massimi sistemi del mondo* in *Opere*, vol. VII, 123, 96-109 ; in the English trans. by S. Drake, *Dialogue on the Two Chief World Systems*. Berkeley, University of California Press, 1962, 97-98, 71-83.

14. *Sidereus Nuncius*, Drake trans., 30 (In Galileo, *Opere*, vol. III, part 1, 68 ; Van Helden trans., 47).

The mention of a geographical area like Bohemia raises an interesting question about the relations between lunar mapping and terrestrial cartography. When drawing parallels between the Moon and the Earth, Galileo did not have a telescopic view of the Earth as a standard for comparison. The only thing that approached such a view was a map or globe, both of which offered artificial projections. Galileo had to bring the Moon down from the sky and project it on a printed page. The transfer had to make the Moon visually comprehensible, and legible on a two-dimensional surface. Galileo's brain helped his eye to draw the Moon with the technique of mapmaking that governed the delineation of such things as coastlines, islands, peninsulas, headlands, basins, and so forth. If we examine more closely the diagrams of the first and last quarters where we found a greatly enlarged crater (Figure 3), we notice that the irregularity of the boundary line is exaggerated, and is traced in a heavily scalloped fashion, as were coastlines in contemporary maps, often with armlike projections reaching out around bays or inlets and with seas shown in darker shading.

The similarity between the play of light and shadow on the Moon and on the Earth is a powerful argument for the existence of mountains and craters on the Moon but it would appear to be subverted by a difficulty that challenges the entire analogy. If the lunar surface is rugged and uneven, why is the circumference of the full Moon perfectly smooth and round ? Galileo offers two answers to this question. The first is correct, the second he was later to repudiate. His first reply consists in acknowledging that if only one range of mountains existed along the extreme edge of the lunar periphery, the Moon would look to us like a toothed wheel. But if a whole series of mountain ranges are closely packed together we could not, from afar, distinguish the separation of prominences by cavities because the spaces between the mountains of a particular chain would be concealed by the mountains of the chain just in front of them.

It is thus that on earth, the summits of several mountains close together appear to be situated in one plane when the observer is a long way off and at an equal height. Likewise, in a rough sea the tops of waves will appear to lie in one plane, though between one high crest and the next there are gulfs and chasms of such depth as to hide not only the hulls, but even the bulwarks, masts, and rigging of stately ships. Now, since there are many chains of mountains and many chasms on the Moon besides those at its periphery, and since the eye, regarding them from a great distance, lies nearly in the plane of the summits, no one need wonder that those appear as if arranged in a regular and unbroken line[15].

Galileo's argument is geometrically compelling as long as the mountain ranges are sufficiently near together for the peaks of one range to block out the

---

15. *Sidereus Nuncius*, Drake trans., 32 (In Galileo, *Opere*, vol. III, part 1, 70 ; Van Helden trans., 49).

dents in the range immediately behind it. But the smoothness of the large
" ancient " spots that Galileo depicts do not, at first sight, make this appear
likely. This is why he produces a second argument that is linked, even more
than the first, with a terrestrial analogy : " There exists around the Moon's
body, just as around the Earth, a globe of some substance denser than the rest
of the aether... which may obstruct our vision, especially when it is lighted,
cloaking the lunar periphery that is exposed to the sun "[16]. By placing what he
terms " a bulk of vapour " around the Moon, on the analogy of the atmosphere
that surrounds the Earth, Galileo went beyond the available evidence. He had
unwittingly stretched the terrestrial comparison to its breaking point. Some-
time after the publication of the *Sidereus Nuncius*, it became clear to him that
there are no clouds on the Moon as should be expected if it had an atmosphere.
Furthermore, since there are only one day and one night during an entire lunar
month, this means that fourteen days of torrid heat are followed by fourteen
days of bitter cold. Hence the impossibility of animal life and even vegetation
on the Moon.

But atmosphere or no atmosphere on the Moon, Galileo was certain of the
existence of mountains and he proceeded to determine their height with the aid
of simple trigonometry. Figure 5 shows a mountain AD whose peak is just
touched by a ray of sunlight GCD. The rest of the mountain still lies in the
dark region beyond the boundary of light CF. From his knowledge of the radius
of the Moon (CE or AE), and his observational determination of the distance
DC, Galileo arrived at the figure of four terrestrial miles for the height of the
mountain AD.

In October 1610, Galileo received a note from the Augsburg scientist
Johann Georg Brengger pointing out that the phenomenon that he recorded
could not have been observed on the rim of the Moon for reasons that Galileo
himself had clearly stated. Namely the rim of the Moon appears perfectly circular, not toothed or dented, because the space between the mountains is concealed by other ranges of mountains. The illuminated spots in the dark region
could only have been observed near the centre. The unevenness of the boundary line between light and darkness made precise measurement impossible, but
it seemed incontrovertible to Brengger that no more than three hours could
have elapsed between the time of the first illumination of a peak in the dark-
ened area and its joining the illuminated boundary. Since the Moon goes
around the Earth (i.e. describes a circle of 360°) in roughly 29 1/2 days, in 3
hours it covers about 1 1/2. This means that the distance CD (see Figure 5) is
much shorter than Galileo claimed and, hence, that the mountain AD need only
be one third of a mile. A mountain four miles high would imply a rotation of
5° and a time of 8 hours, much more than Galileo had intimated.

16. *Sidereus Nuncius*, Drake trans., 32-33 (In Galileo, *Opere*, vol. III, part 1, 70 ; Van Helden
trans., 49-50).

In a lengthy reply, which is one of the first detailed discussions of the application of geometry to the new celestial data, Galileo granted that Brengger's reasoning was valid but claimed that some peaks are indeed illuminated more than eight hours beyond reaching the boundary of light. All that could be concluded was that mountains on the Moon are of varying heights ! More interesting, perhaps, is Galileo's avowal that his data were taken from the central part of the Moon. He had to draw the mountain as though it were on the very rim of the Moon in order to make his point. The geometrical diagram is turned through a right angle and shows the Moon's axis, so to speak, pointed toward our eyes, while the Moon's equator is shown as its rim. This did not entirely settle the issue of the actual heights of the mountains, as Brengger firmly but courteously remarked in a second letter : the bright summits that gradually reach the boundary line can hardly be illuminated for more than three hours, not the eight an half hours that Galileo claimed. This letter seems to have gone unanswered[17].

Galileo mapped the Moon in an age when " new worlds " were being discovered on Earth as well. Explorers venturing into North and South America tended to see what they expected to find, and their reports are full of comparisons with what was already familiar to them either from personal experience or from reading and conversation. Galileo was conscious of the role of preconceived ideas on the way we shape our sense data. In the First Day of the *Dialogue on the Two Chief World Systems*, he shows how our imagination, however bold and playful, is limited by everyday experience. He asks us to consider the case of a person who is born in a large forest where there are birds and wild beasts but no aquatic animals. Since such a person would never have seen a lake or a stream, he could know nothing about fishes. He could not possibly imagine animals that move without walking on their legs or beating their wings, or are able to go up and down or even stop motionless, something that birds in the air cannot do. Neither could he imagine that human beings like himself build vessels that enable them to move their entire household very easily from one place to another. " Even with the liveliest imagination, such a man could never picture to himself fishes, the ocean, ships, fleets, and armadas "[18].

NEW EYES FOR A NEW AGE

Galileo's own lively imagination was sustained by the telescope, but the new instrument could only perform marvels in the hands of someone who was prepared to transform the lunar surface into a terrestrial landscape. That this is

17. Brengger's query was communicated to Galileo by Marc Welser on 29 October 1610, *Opere*, vol. X, 460-462. Galileo replied on 8 November 1610, *Opere*, vol. X, 466-473, and Brengger urged his objection in a letter dated 13 June 1611, *Opere*, vol. XI, 121-125.

18. Galileo, *Dialogue on the Two Chief World Systems*, *Opere*, vol. VII, 86 ; Drake translation, 61.

the case can be seen from the way another great scientist looked at the Moon several months before Galileo thought of doing so. On 26 July 1609, the English mathematician, cartographer and astronomer Thomas Harriot became the first person to produce an image of the Moon on the basis of telescopic observations. With a 6-power telescope, he drew a five-day-old Moon[19]. As Figure 6 shows, this is a rather crude rendering of the crescent phase with an irregular boundary line and darkened patches that were later called maria. It is more of a sketch than a drawing and it lacks any written comment. Harriot attached no great importance to what he had recorded since he made no further observations in the days that followed. He had not expected to see a world on the Moon, and he did not find what he had not been searching for. A year later, after receiving a copy of Galileo's *Sidereus Nuncius*, his interest was rekindled and he had a fresh look at the Moon. The very first of his new drawings (Figure 7), with its perfectly circular crater at the center, reveals how fast he was learning to observe with Galileo's eye (compare with Figure 4). Too fast perhaps, but once given a clue, Harriot looked for more all over the Moon. On 26 August, he had found valleys and, by 11 September, islands and promontories.

Such a quest was not easy. Before seeing Galileo's diagrams, Harriot had sent his first telescope to a friend, Sir William Lower, who reported on his observations with " the perception cylinder " on 6 February 1610 : " According as you wished I have observed the moone in all his changes... [Near] the brimme of the gibbous part towards the upper corner appeare luminous parts like starres, much brighter than the rest, and the whole brimme along lookes like unto the description of coasts, in the dutch bookes of voyages. In the full she appeares like a tarte that my cooke made me the last weeke "[20].

Lower is groping for an apt description of what he sees. Words fail him, and his imagination is tossed from the description of a coastline read in a Dutch travel-book to the memory of last week's pie. The terrestrial features of the Moon seem to cry out to be recognised but Lower's vision is both overwhelmed and blurred. The analogy that removed all confusion was conveyed to him by Harriot in the Summer of 1611, and Lower immediately wrote back : " Me thinks my diligent Galileus hath done more in threefold discoverie than Magellane in openinge the streights to the South Sea or the dutchmen that were eaten by the beares in Nova Zembla. I am sure with more ease and safetie to him selfe & more pleasure to mee. I am so affected with newes as I wish sommer were past that I mighte observe the phenomenes also. In the moone I had

19. Quoted in J.W. Shirley, " Thomas Harriot's Lunar Observations ", in *Science and History : Studies in Honor of Edward Rosen. Studia Copernicana*, VII (1977), 303. See also T.F. Bloom, " Borrowed Perceptions : Harriot's Maps of the Moon ", *Journal for the History of Astronomy* , IX (1978), 117-122, and S.L. Montgomery, *The Scientific Voice*, New York, London, The Guilford Press, 1996, 212-219.

20. Quoted in E.A. Whitaker, " Selenography in the Seventeenth Century ", 120.

formerlie observed a strange spottednesse al over, but had no conceite that anie parte thereof mighte be shadowes "[21].

Once the significance of shadows is clear, the geographic metaphor, once glimpsed but discarded, is immediately reinstated. Galileo is hailed as a great explorer, and the Moon is described as a new land to be crossed, charted and colonized. When Galileo visited Rome in the Summer of 1611 he was given a hero's welcome, and Jesuits and Church dignitaries vied in their praise. On 31 May, Cardinal Francesco Maria del Monte wrote to the Granduke of Tuscany, Cosimo II, " If we lived in ancient Rome, I am certain that a statue would have been erected to him on the Campidoglio "[22]. No mild praise when we think that the Campidoglio was dominated by the equestrian figure of the Emperor Marcus Aurelius !

An equally fitting tribute to Galileo's daring use of the science of perspective came from his artist friend Cigoli, who was working at the time in the Pauline Chapel of the Basilica of Santa Maria Maggiore in Rome. Cigoli had been commissioned to paint the vision of " a woman clothed with the sun, with the Moon under the feet and a crown of twelve stars on her head " (*Revelation*, ch. 12, v. 1). The representation of the woman is conventional enough, but the Moon is depicted in an entirely novel way, as a crater-pocked crescent. Galileo had brought the Moon down to Earth with his telescope ; with his brush, Cigoli triumphantly restored it to the heavens[23].

CONCLUSION

We see what we look for, and our vision is guided by our expectations. Plutarch had speculated that the Moon might be a planet similar to the Earth and, had Galileo had not been aware of this hypothesis, he might not have seen mountains and valleys when he focused his primitive telescope on our satellite. As the reader who as glanced through a telescope will testify, the Moon hardly looks like the Earth. Its bleached or blackened surface, pockmarked with circular craters and mottled throughout, does not resemble any picture taken on our planet. The modern Moon had to be invented, and this was rendered possible by comparing the progress of sunrise on Earth with the play of light and shadow on the Moon.

In other words, the lunar surface had to be interpreted as a landscape. The four drawings that Galileo chose to be engraved for the *Sidereus Nuncius* show quarter and half-moon phases because these, and not an image of the full

21. Quoted *Ibid.*, 120-121.

22. Galileo, *Opere*, vol. XI, 119.

23. See F. Hallyn, " Pensée religieuse et pensée scientifique : une fresque Galiléenne ", dans G. Mathieu-Castellani (éd.), *La pensée de l'image*, Paris, l'Imaginaire du texte, 1977, 51-61. Hallyn shows that the painting, which has often been assumed to represent the Assumption of Mary or even the Immaculate Conception, is really a depiction of the vision in the book of *Revelation*.

Moon, reveal how landforms come into view across the dawning surface. Harriot had to be helped to see this, but once his eyes had been opened, he experienced no difficulty in recognising promontories and circular cavities. The Moon as we see it today is the happy outcome of the combination of an old hypothesis and a new device that began its life as a toy. The hypothesis turned the toy into a scientific instrument, and together they transformed a perfectly spherical globe into the rugged and uneven surface on which twentieth-century man would eventually land.

FIGURES

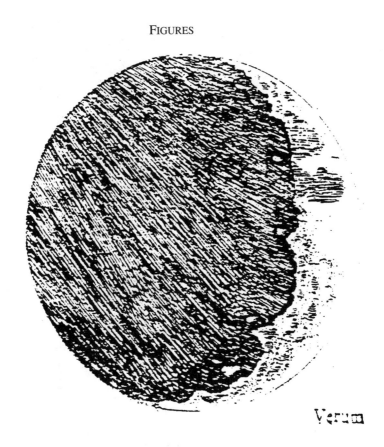

Verum

1. The Moon four or five days old as drawn by Galileo.

2. Engravings of the Moon just before and at the last quarter
from Galileo's Sidereus Nuncius.

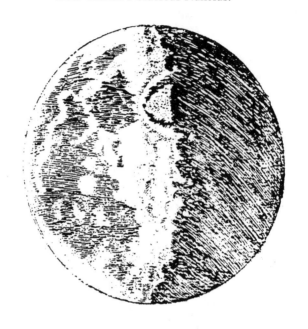

3. Engravings of the first and last quarter of the Moon
from Galileo's *Sidereus Nuncius*.

OBSERVAT. SIDEREAE

dum daturam. Depressiores insuper in Luna cernuntur magnæ maculæ, quàm clariores plagæ; in illa enim tam crescente, quam decrescente semper in lucis tenebrarumque confinio, prominente hincinde circa ipsas magnas maculas contermini partis lucidioris; veluti in describendis figuris obseruauimus; neque depressiores tantummodo sunt dictarum macularum termini, sed æquabiliores, nec rugis, aut asperitatibus interrupti. Lucidior verò pars maximè propè maculas eminet; adeò vt, & ante quadraturam primam, & in ipsa fermè secunda circa maculam quandam, superiorem, borealem nempè Lunę plagam occupantem valdè attollantur tam supra illam, quàm infra ingentes quædam eminentiæ, veluti appositæ præseferunt delineationes.

Hæc

RECENS HABITÆ 10

Hæc eadem macula ante secundam quadraturam nigrioribus quibusdam terminis circumuallata conspicitur; qui tanquam altissima montium iuga ex parte Soli auersa obscuriores apparent, quà verò Solem respiciunt lucidiores extant; cuius oppositum in cauitatibus accidit, quarum pars Soli auersa splendens apparet, obscura verò, ac vmbrosa, quæ x parte Solis sita est. Imminuta deinde luminosi aperturæ, cum primum tota fermè dicta macula tenebris est obducta, clariora mōtium dorsa eminenter tenebras scandunt. Hanc duplicem apparentiam sequentes figuræ commōstrant.

C 2 Vnum

4. Moon at last quarter as seen through a modern telescope and as drawn by
Galileo. From S. Drake, *Galileo at Work*, Univ. of Chicago Press, 1978, 145.

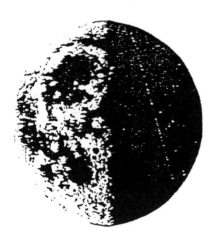

5. Galileo's diagram of the Moon showing the height of mountain AD.

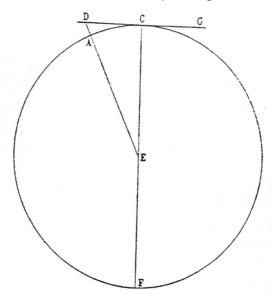

6. Harriot's first lunar drawing, dated 26 July 1609. Adapted from T.F. Bloom, " Harriot's Maps of the Moon ", *JHA*, IX (1978), 118.

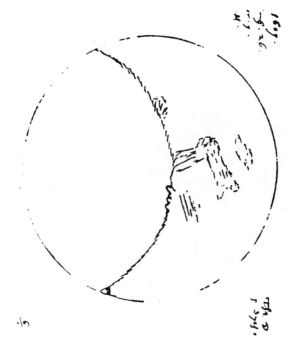

7. Harriot's second lunar drawing, dated 17 July 1610. From T.F. Bloom,
   " Harriot's Maps of the Moon ", *JHA*, ix (1978), 119.

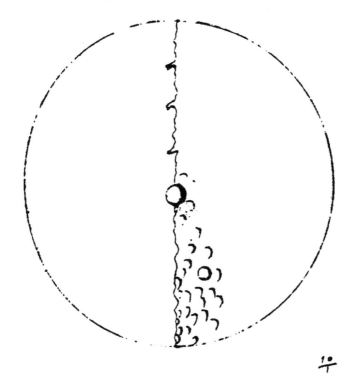

Réflexions sur " l'optique " de Galilée

Philippe Hamou

Bien que Galilée ait plusieurs fois annoncé son intention d'écrire sur des sujets optiques[1], il n'a jamais réalisé ce projet. Les témoignages textuels de son savoir optique sont rares, allusifs et dispersés. Ils coexistent avec des déclarations d'ignorance ou d'incompréhension sur la nature de la vision, l'essence de la lumière ou la science des réfractions[2]. Vasco Ronchi estimait que, " contrairement à l'opinion générale, Galilée n'a jamais poussé à fond ses recherches d'optique "[3]. L'incompétence optique de Galilée, en particulier pour tout ce qui touche à la réfraction, a fait l'objet de remarques similaires chez Santillana, Geymonat, Shea ou récemment Pantin, et, s'il existe bien, dans la littérature secondaire, quelques appréciations positives de l'optique de Galilée, elles sont le plus souvent énoncées de manière péremptoire[4] ou étayées par des hypothèses fragiles et invérifiables, comme celle avancée par Machamer selon laquelle Galilée aurait eu accès, bien avant sa publication en 1611, au manuscrit des *Photismi* de Maurolyco, y puisant le savoir sur les propriétés des lentilles nécessaire à la construction de sa lunette[5].

Il n'entre pas dans mon propos de tenter ici une reconstitution et une critique de ce que pouvait être le savoir optique de Galilée et des sources auxquelles il s'alimentait. Mon intention est d'offrir plutôt quelques éléments pour une

1. *Cf. Sidereus Nuncius*, éd. I. Pantin, Paris, Les Belles Lettres, 1992, 9 et la lettre à Vinta de 1611, in *Le Opere di Galileo Galilei*, Edizione Nazionale, éd. A. Favaro, Florence, 1890-1909 (cité EN), vol. X, 352.
2. *Cf.* notamment EN, VI, 350 où Galilée avoue " ne comprendre que bien peu de choses " au sujet de la vision ; la lettre à Liceti du 23 juin 1640 (EN, XVIII, 208), où il déclare " avoir toujours été dans les ténèbres " sur l'essence de la lumière.
3. V. Ronchi, *Histoire de la Lumière*, trad. fr. J. Taton, Paris, Armand Colin, 1956, 84.
4. *Cf.* par exemple A. Koyré qui affirme, mais sans citer de documents, que " le télescope galiléen n'est pas un simple perfectionnement de la lunette batave, il est construit à partir d'une théorie optique " (" L'apport scientifique de la Renaissance ", in *Etudes d'Histoire et de Philosophie des Sciences*, Paris, Gallimard, 1973, 59).
5. P.K. Machamer : " Feyarabend and Galileo : the interaction of theory, and the reinterpretation of experience ", *Studies in History and Philosophy of Science*, 4, n°1 (1973).

élucidation de " l'idée galiléenne de l'optique ", le statut qu'il accordait à cette science au sein des disciplines mathématiques et physiques, sa fonction pour l'interprétation de la nature. Je tenterai sur ces bases d'esquisser une réponse à une question lancinante : que signifie l'espèce de désinvolture théorique dont il fit preuve au sujet du fonctionnement optique du télescope ?

L'IDÉE GALILÉENNE DE L'OPTIQUE

Il convient certainement de distinguer deux dimensions de la science optique, à l'égard desquelles l'attitude de Galilée est très différente. Il y a d'une part l'optique ou la *prospettiva* euclidienne, une de ces sciences mathématiques mixtes comme la musique ou l'astronomie, qui, pour le dire dans les termes de Galilée lui-même, naissent lorsque les mathématiciens se mettent à faire des observations sensibles. Traitant du rapport entre ce que les choses sont géométriquement et la manière dont elles apparaissent à la vision directe, c'est une discipline dont la base expérimentale est extrêmement réduite (limitée à quelques brefs axiomes), et qui peut apparaître comme étant entièrement l'affaire du mathématicien. Galilée, qui fut professeur de mathématiques à Pise et à Padoue, et qui y a peut-être enseigné la " perspective ", se reconnaît une pleine compétence dans ce domaine — compétence attestée par les témoignages concordants de Cigoli et Viviani, mais surtout par l'usage constant que Galilée fait dans son œuvre astronomique des théorèmes et des concepts de l'optique euclidienne, notamment lorsqu'il s'agit de déterminer une parallaxe, interpréter le relief lunaire, calculer la hauteur des montagnes de la lune ou encore déduire de la configuration et des raccourcis des taches solaires leur vraie forme et situation.

Il existe par ailleurs dans l'œuvre de Galilée une série de recherches qui méritent tout aussi bien la qualification d'optique, mais qui sont bien loin de posséder le caractère de science certaine et close revêtu par la perspective. Ce sont des recherches de nature expérimentale touchant la réflexion et la réfraction, les illusions d'optiques qu'elles entraînent, la nature de la vision et de la lumière. Il est clair qu'il existe à l'époque de Galilée sur ces questions de nature plus " physique " une littérature optique au moins aussi abondante que celle consacrée à la *prospettiva*, mais Galilée a très tôt exprimé sa réticence à l'égard de cet héritage, caressant le projet de reprendre ces questions *ab initio*, sur la base de recherches expérimentales et mathématiques nouvelles. On trouve dès 1590 des considérations sur le fait que les phénomènes de la réfraction et de la réflexion ne sont pas encore bien connus, et qu'à leur égard, comme à l'égard du mouvement, on se fie à tort à des opinions enracinées plutôt qu'à l'expérience attentive[6]. Les considérations sur la réfraction dans

---

6. *Cf.* le passage consacré à l'apparence d'un denier plongé dans l'eau dans le manuscrit du *De Motu* de 1590 édité par I.E. Drabkin in *Galileo Galilei on Motion and Mechanics*, Madison, 1960, 83-84.

L'Essayeur, celles consacrées aux phénomènes catoptriques dans la Première Journée du *Dialogue* révèlent une attitude similaire. Loin de penser que la littérature optique traditionnelle (Alhazen et Vitellion) dont ses adversaires se réclament a recensé et expliqué tous les phénomènes particuliers de l'optique physique, Galilée estime d'une part qu'il faut réapprendre à voir les phénomènes élémentaires (qu'on croit faussement bien connaître), comme celui des images sur les miroirs sphériques, et d'autre part qu'il reste beaucoup de phénomènes à découvrir, " la nature produisant ses effets par des voies auxquelles nous ne penserions point si les sens et l'expérience ne nous les montraient, ce qui parfois d'ailleurs ne suffit pas à suppléer à notre incapacité "[7].

Cette manière de considérer l'optique " physique " comme une science encore inchoative explique deux traits particuliers dans les recherches de Galilée : d'une part son attention aux micro-phénomènes optiques, généralement négligés ou passés sous silence par la tradition : scintillations, halos, réflexions multiples, franges colorées, lumière cendrée de la Lune, *etc.*, attention aux données fines de l'expérience visuelle qu'on ne retrouvera peut-être pas à ce degré avant Grimaldi, Hooke ou Newton ; d'autre part son manque manifeste d'intérêt à l'égard de ce que la tradition ou la littérature optique du moment pouvait lui apporter de solide sur les questions optiques fondamentales comme la propagation lumineuse, la réfraction ou la physiologie oculaire. Pour une science comme l'optique, où la continuité historique fut peut-être plus remarquable encore que les ruptures, cette négligence était funeste. Elle explique sans les excuser les cafouillages assez grossiers de Galilée sur la réfraction atmosphérique[8], ou le fait qu'en 1632, dans le *Dialogue,* il n'ait toujours pas pris acte de la réforme képlérienne du *modus visionis.*

Il existe, entre les deux dimensions de la science optique que Galilée considère, une différence importante en ce que la première est d'ores et déjà mathématisée, tandis que la seconde est en instance de mathématisation. La perspective euclidienne pouvait ainsi sans attendre servir la fin que Galilée assignait aux mathématiques mixtes : géométriser la nature, pour raisonner enfin avec la certitude du mathématicien dans des questions de philosophie naturelle.

L'attitude philosophique traditionnelle (aristotélicienne) à l'égard des sciences mathématiques mixtes comme l'astronomie ou l'optique était de considérer que leur certitude géométrique est aussi la raison pour laquelle elle ne permettent pas de bien raisonner physiquement. L'opticien ou l'astronome ne peut prétendre tirer des conclusions physiques de ses arguments mathématiques parce qu'il raisonne sur des objets (les apparences sensibles indistinctes des astres) qui n'ont pas l'exactitude des objets mathématiques. " Il n'est point né

7. *Cf. L'Essayeur de Galilée*, trad. Chauviré, Paris, Les belles Lettres, 1980, 183 ; EN, VI, 281.
8. *Ibid.*, 244-245 ; EN, VI, 354.

capable de contempler la cause "[9] des phénomènes et doit être instruit par celui qui tire des premiers principes de la physique ce qui touche à l'essence du ciel et des astres. Cette doctrine commençait à l'époque de Galilée à être sérieusement contestée. La préface que Jean de la Pène (Johannis Pena) joignit à sa traduction latine de l'*Optique* d'Euclide (1557) contient ainsi un intéressant " éloge de l'optique " où il est dit en substance que l'optique n'a pas seulement pour fonction d'être utile aux praticiens comme les peintres ou les arpenteurs, mais qu'elle a aussi un rôle important à jouer au service des philosophes. Par elle, on peut prouver (en faisant usage des parallaxes) la fausseté de l'argument copernicien sur le mouvement de la terre, ou encore le caractère supra-lunaire des comètes. Il est intéressant de noter que Kepler consacra une bonne partie de la préface de sa propre *Dioptrique* (1611) à discuter les arguments de Pena, dont il accepte le principe (la valeur philosophique de l'optique) mais non évidemment l'application particulière qui en est faite s'agissant du copernicanisme : " ce n'est pas faire valoir l'optique que de solliciter ses forces pour des choses impossibles ".

L'usage que Galilée réserve à l'optique géométrique s'inscrit incontestablement dans ce mouvement de réévaluation philosophique des sciences mathématiques, ce travail de sape du territoire réservé de la philosophie naturelle. Cela est manifeste dès 1605 dans le débat qui l'oppose aux aristotéliciens de Padoue sur l'interprétation de l'étoile nouvelle de 1604[10] ; et plus explicitement en 1611, lorsque, à l'occasion de la bataille du *Sidereus Nuncius*, il doit défendre contre La Galla l'usage d'arguments optiques empruntés à la perspective des ombres pour démontrer la qualité terrestre (rugueuse, montagneuse) de la Lune. La Galla considère que " même lorsque les mathématiques s'abaissent sur des quantités physiques, comme c'est le cas en optique et en harmonique, rien dans ces contemplations ne concerne les qualités physiques et naturelles, qualités que les mathématiques considèrent toujours comme des accidents "[11]. Galilée annotant l'ouvrage de son adversaire rétorque : " il n'est pas moins ridicule de dire que le géométrique ne correspond pas au matériel que de dire que les propriétés arithmétiques dans les corps ne correspondent pas aux nombres — et que par exemple les règles pour ordonner et coordonner les armées ne valent pas lorsque nous avons affaire à de vrais corps militaires "[12]. La conséquence absurde de l'argument de La Galla est qu'il vaudrait mieux se crever les yeux pour faire de la bonne philosophie. La suite du texte de La Galla se concentre sur les illusions optiques dont Galilée a pu être victime dans sa description de la Lune. En particulier la Galla évoque le fait qu'une peinture plate

---

9. *Cf*. Simplicius citant Geminus dans ses Commentaires sur la *Physique* d'Aristote. Texte traduit par P. Duhem, in ΣΩZEIN TA ΦAINOMENA, Paris, Vrin, 1983, 7-9.

10. *Cf*. les documents réunis par Drake in *Galileo against the Philosophers*, Los Angeles, Zeitlin & Ver Brugge, 1976.

11. *Cf*. J.C. La Galla, *De Phaenomenis in Orbe Lunae*, Venise, 1612 reproduit in EN III-I, 323.

12. EN III-I, 323.

peut apparaître exhiber des proéminences et des rugosités qui n'existent pas. Galilée répond : " si ces illusions sont le fait de la perspective, qui peut le mieux les corriger et les comprendre que les perspectivistes eux-mêmes ? " (...) " Vous voulez accuser les mathématiciens d'ignorer que les sens nous trompent dans les choses perçues communément, comme si le fait de savoir si l'on est trompé ou pas était un mystère philosophique des plus profonds et secrets. Mais quels sont ceux qui ont fait les plus exactes et les plus nombreuses observations et théories sur le sujet des erreurs de la vision que ces mêmes mathématiciens ? "[13].

Il est clair que Galilée revendique une autorité complète sur un domaine qui n'est pas seulement celui de la pure mathématique, mais de la mathématique jointe à l'observation. Non seulement le mathématicien est capable d'appliquer une analyse quantitative à l'observation mais il est également capable d'apprécier le plus et le moins dans cette analyse, le degré d'approximation auquel il peut prétendre et la nature des illusions ou des erreurs perceptives qu'il lui faut surmonter. Ce faisant Galilée attribue au mathématicien (ici le perspectiviste) la compétence non seulement pour juger de la validité d'un raisonnement — mais — ce qui est plus déconcertant pour un aristotélicien orthodoxe — la compétence également pour juger de la validité d'une expérience sensible appelée à fonder un raisonnement physique.

L'OPTIQUE TÉLESCOPIQUE DE GALILÉE

Le fonctionnement optique du télescope manifeste assurément l'interpénétration des deux champs optiques que Galilée tendait à maintenir séparés — celui de l'expérience stable, rigidement mathématisable, ouvert par la *prospettiva* et celui de l'expérience multiforme encore mal comprise et mal théorisée associée à la réfraction et à la physiologie optique. Il est clair qu'au regard de la science dioptrique ultérieure, celle qui se développe à partir de Kepler et Descartes, la simple perspective euclidienne n'est pas suffisante pour rendre compte des effets de l'instrument. Galilée, qui ne l'ignore pas, a pourtant une attitude paradoxale : tout son effort est de montrer que le télescope fonctionne *mutatis mutandis* comme un opérateur euclidien dont l'effet est d'ouvrir les pyramides visuelles sous-tendues par les objets, et de produire ainsi au regard un visible bien constitué sur lequel peut s'exercer à bon droit le jugement des sens. A cette fin il développe une optique *ad hoc* qu'on pourrait bien juger assez sévèrement et qui est sans doute plus révélatrice des intentions épistémologiques de Galilée que de la nature réelle de l'instrument.

Galilée a toujours affirmé être parvenu à construire la lunette *per raggione* et en recourant en la doctrine des réfractions. Mais les textes de l'*Essayeur*[14]

---

13. EN, III, 394-397.
14. *L'Essayeur*, 164-165, EN VI, 258-259.

où il livre son raisonnement ne laissent guère de doute sur le type de savoir mis en jeu : c'est un savoir empirique de lunetier dont on pouvait déjà trouver un énoncé précis dans les traités pratiques de la fin du XVI siècle : on s'intéresse aux propriétés réfractives de chaque lentille, puis à la composition de ces propriétés, et non pas au système optique que constitue l'instrument. Ces propriétés sont conçues comme affectant de manière additive des caractéristiques phénoménales qualitatives (grandeur et petitesse, confusion et distinction) et non pas des entités abstraites extra-phénoménales (la trajectoire des rayons, la distance focale, le point image). Il est assez peu vraisemblable que Galilée, qui ne recourt jamais à la notion de convergence des faisceaux de rayons en un point, ait construit ses instruments plus perfectionnés en faisant appel, comme on l'a parfois suggéré, à un calcul de distance focale. Il semble bien plus probable que le perfectionnement du télescope s'est fait *a posteriori*, à partir des effets constatés sur une série de lentilles étalons.

Dans l'interprétation optique qu'il essaya par la suite de donner de ces effets Galilée ne cherche pas une théorie unifiée et solide, mais plutôt une série d'arguments *ad hoc* qui lui permettraient de dire que la vision dans le télescope livre une expérience authentique et stable, et que l'on a donc toutes les raisons de s'appuyer sur elle pour construire des raisonnements physiques et astronomiques.

Ainsi, dans le *Sidereus Nuncius* où Galilée donne un schéma grossier du tube optique et une explication des effets de réduction de champ de la lunette associé à l'usage des lentilles, il est frappant de constater qu'il ne cherche pas à représenter les effets particuliers et pour lui encore obscurs de la réfraction dans l'instrument. La lunette est prise comme un tout, une boîte noire ou un opérateur mathématique qui transforme l'angle visuel EFG défini par la vision à travers le tube vide ABCD en l'angle plus resserré EHI, chose dont on peut s'assurer par des observations répétées sur des objets terrestres accessibles. Cette description euclidienne se fait au prix du gommage pur et simple des phénomènes qui ne répondent pas adéquatement à la reconstruction géométrique : ainsi, Galilée, qui avait très certainement compris par expérience que la variation de l'ouverture de l'objectif pour une lunette de dimension donnée n'entraîne pas une variation proportionnelle du champ de la lunette, ne mentionne pas ce point mais décrit plutôt un procédé impraticable pour mesurer les distances angulaires par l'usage de cartons découpés placés sur l'objectif de la lunette[15].

Le fait que Galilée tend à considérer son instrument comme une sorte d'épure géométrique trouve une confirmation paradoxale dans le traitement d'un effet annexe du télescope. Galilée dès le *Sidereus Nuncius* avait constaté que les planètes et les étoiles perdent au télescope tout ou partie de leur " chevelure ", et tendent à se résoudre en de petits disques nettement délimités.

---

15. *Cf. Sidereus Nuncius*, 8-9.

Dans l'*Essayeur*, il attribue cette irradiation trompeuse à un effet de la réflexion sur le bord humide des paupières : " cet éclat adventice des étoiles, explique-t-il, n'est pas en réalité autour des étoiles mais dans notre œil, de l'étoile ne parvient que sa simple *specie* nue et nettement délimitée "[16]. Le télescope en grossissant les objets et en concentrant la lumière du milieu rend imperceptible l'effet d'irradiation qui, étant le fait de l'œil, ne traverse pas les lentilles et n'est donc pas accru. Ici encore, au terme d'un raisonnement passablement sinueux, la lunette est présentée comme un instrument parfait, qui remédie au défaut de l'œil en permettant à l'image des objets de lui parvenir telles qu'elles sont en soi, à savoir " nues et bien délimitées ", avec la précision géométrique requise pour l'application pertinente de l'optique à l'astronomie. La suppression télescopique de l'irradiation offre en effet, rappelons-le, un remarquable argument pro-copernicien. L'observation à l'œil nu de Mars et de Vénus, affectée par l'irradiation ne laissait voir que très peu de changement de magnitude entre les positions de conjonction et d'opposition, ce qui paraissait contredire sensiblement le système de Copernic en vertu duquel les deux disques, en rotation autour du soleil devaient présenter des variations de diamètre apparent de l'ordre de 1 à 40 ou 1 à 60. Grâce à " une lunette parfaite ", comme le dit Galilée, on peut désormais voir cette variation et comme " la toucher du doigt "[17].

Concluons sur l'analyse optique de l'instrument offerte par Galilée : elle est volontairement elliptique, souvent *ad hoc* et ne semble pas manifester une grande cohérence (rien par exemple ne vient unifier les propriétés d'accroissement et de distinction de la lunette). Toutefois il y a un fort élément unifiant dans tout ce que Galilée dit au sujet de la lunette : son but est de montrer que le visible qu'elle offre est éminemment susceptible d'analyse géométrique ou perspective. La lunette en effet permet à la géométrie perspective de s'exercer là où l'œil nu ne le peut pas, en offrant une vision plus performante dans les lointains, et une vision épurée des effets adventices de la lumière et de l'œil. On voit bien sur l'exemple des taches solaires ou des montagnes de la Lune que Galilée peut désormais, grâce à la lunette, appliquer aux objets astronomiques ces raisonnements optiques ou perspectifs que les aristotéliciens estimaient inopérants en raison de la trop grande distance. Comprise de cette manière, la lunette est bien l'une des armes dont Galilée avait besoin dans son entreprise fondamentale de réappropriation complète du champ de la physique par le raisonnement mathématique et par l'expérience sensible bien constituée. On ne s'étonnera donc pas que la lunette ne fut pas pour Galilée ce qu'elle fut pour Kepler et Descartes : une provocation épistémologique qui invitait à repenser les rapports de l'œil et du monde. Si, chez Galilée la lunette n'assigne aucune signification nouvelle au regard qui la traverse, c'est parce que pour

16. *Ibid.*
17. *Ibid.*, 141-142, EN VI, 232-233.

livrer au regard les choses mêmes, le monde physique et réel des astres, elle doit être de part en part transparente, n'ajouter rien de son fait à la parfaite géométrie des pyramides visuelles. Aussi, elle fut bien pour Galilée ce *perspecillium*, cet objet transparent au travers duquel on voit mais qui, semblable au panneau de verre des peintres perspectivistes, n'est lui-même jamais vu et ne doit jamais l'être.

DE KEPLER À DESCARTES : LE RENOUVELLEMENT DE LA PROBLÉMATIQUE DE LA VISION

Gérard SIMON

La découverte de l'image rétinienne par Kepler, déplaçant du cristallin à la rétine le premier siège de la sensibilité oculaire, n'a pas été un simple avatar couronnant les efforts des " perspectivistes " médiévaux ; elle a au contraire induit une véritable crise dans la conception traditionnelle de la vision, conduisant à une très profonde restructuration opérée par Descartes dans sa *Dioptrique*. Pour être bref, à la perplexité de Kepler devant sa propre découverte, Descartes a répondu en distinguant clairement une phase lumineuse, une phase nerveuse et une phase mentale de la perception visuelle, renonçant du même coup à postuler, comme le faisait toute la tradition optique, une nécessaire ressemblance entre l'impression psychique ressentie et sa cause physique externe. Ce qui n'alla pas sans conséquences épistémologiques, anthropologiques et philosophiques.

LA PERPLEXITÉ DE KEPLER

Quand en 1604 Kepler eût démontré géométriquement dans ses *Ad Vitellionem Paralipomena, seu Astronomiae Pars Optica* que les faisceaux lumineux issus de chaque point de l'objet et entrés dans l'œil par la pupille convergent grâce au cristallin en un point unique de la rétine, formant sur cette dernière une " peinture " (nous dirions une image réelle) de l'objet, il fut bien embarrassé. Car il modifiait la fonction attribuée avant lui au cristallin, qui jouait dans sa conception le rôle d'une simple lentille convergente : le cristallin cessait donc d'être le lieu où les rayons lumineux venus de l'extérieur suscitent la formation d'une image sensorielle transmise à travers le nerf optique et le chiasma jusqu'aux facultés supérieures sises dans le cerveau, comme le pensaient Alhazen avec à sa suite Vitellion et tous les perspectivistes médiévaux. Or Kepler quant à lui, pour expliquer comment l'image de l'objet passe de la rétine au cerveau, ne trouvait rien de satisfaisant à proposer : " Je dis que la vision se produit, quand l'image (*idolum*) de la partie hémisphérique du monde située devant l'œil, et même d'un peu plus, se forme sur la paroi blanc rougeâtre de la surface concave de la rétine. Comment cette image ou cette peinture (*pictura*) se lie aux esprits visuels qui résident dans la rétine et dans le nerf ; savoir si c'est par ces esprits qu'elle est amenée à travers les cavités du cerveau devant le tribunal de l'âme ou de la faculté visuelle, ou si au contraire c'est la faculté visuelle qui, comme un questeur délégué par l'âme, descendant du prétoire du cerveau jusque dans le nerf optique et la rétine comme jusqu'à ses derniers bancs, s'avance au devant de cette image, cela, dis-je, je laisse aux physiciens le soin d'en décider "[1]. A deux reprises encore, et dans des termes voisins, Kepler répéta l'expression de son embarras : en 1611, dans sa *Dioptrice*, où il va jusqu'à remettre en question le rôle transmetteur du nerf[2], et à nouveau en 1619 dans le livre IV de son *Harmonice Mundi*[3], où il affirme être moins démuni pour expliquer la perception de la position relative des astres errants par l'âme de la Terre (admise dans ses conceptions astrologiques) que pour comprendre comment se parachève la vision des choses externes par l'âme de l'homme.

Il fallait pourtant être Kepler pour réaliser que sa découverte rendait caduque la solution des perspectivistes, dont nous ne pouvons donner ici qu'une esquisse synthétique[4]. Ces derniers, quelles que soient leurs différences, fondaient leur explication sur la transmission jusqu'au cerveau d'une quasi-image qu'ils pensaient se former sur le cristallin (fig. 1). Parmi tous les rayons lumineux renvoyés de chaque point de l'objet à l'œil, ils prenaient prioritairement en compte celui qui vient frapper à la perpendiculaire la cornée et la face antérieure du cristallin, supposées concentriques, sans donc subir de réfraction. Ainsi à un point de l'objet correspondait sur le cristallin un point coloré et un seul, et tous ces points donnaient au total une quasi-image de l'objet. Au niveau du cristallin, cette quasi-image lumineuse devenait une image sensorielle dont la transmission ultérieure, bien que convoyée par des " esprits " visuels et non plus par des " espèces " lumineuses, était conçue sur un modèle optique. Les esprits visuels se propageaient de manière rectiligne à travers le cristallin, et subissaient sur sa face postérieure une réfraction qui les envoyait en un faisceau parallèle dans le nerf optique, supposé creux et dans l'axe de l'œil, sans que l'image fût déformée ou inversée. Arrivées au chiasma, les images de chaque œil fusionnaient pour donner une image unique qui passait jusqu'au cerveau, où elle était saisie par le sens commun, jugée par l'intellect, mise en mémoire, *etc.*

1. J. Kepler, *Ad Vitellionem Paralipomena*, livre V, 2. *Gesammelte Werke* (ci-dessous *G.W.*) , München, Beck, tome 2, 151, (traduit par nous).

2. J. Kepler, *Dioptrice*, prop. LXI, *G.W.*, t. 4, 373.

3. J. Kepler, *Harmonice Mundi*, livre IV, *G.W.*, t. 6, 274.

4. Voir G. Simon, " La théorie cartésienne de la vision, réponse à Kepler et rupture avec la problématique médiévale ", *Descartes et le Moyen Age*, Paris, Vrin, 1997, 109-111.

Figure 1 : La vision selon Vitellion
(*Cf. Vitellonis Opticae*, III, 17, *Opticae Thesaurus*, Bâle, 1572, p. 92)

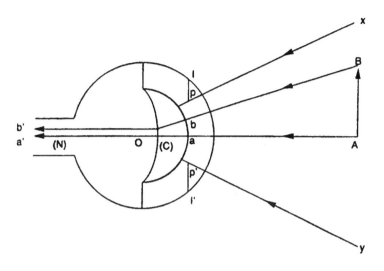

AB	objet	(C)	cristallin
ab	quasi-image cristallinienne	O	centre de l'œil
pp'	pupille	(N)	nerf optique
II'	iris	a'b'	forme sensorielle

- Aa, Bb, *etc.* rayons lumineux utiles (normaux à la cornée et à la face antérieure du cristallin).
- a'b' forme ou image sensorielle de l'objet se propageant dans l'humeur vitrée et le nerf optique.
La face postérieure du cristallin pourrait aussi être plane ou légèrement convexe. Mais toujours située dans la partie antérieure du globe oculaire, elle transmet par réfraction les formes parallèlement à l'axe de l'œil.

Ce qui fait difficulté pour Kepler, c'est l'usage métaphorique de la propagation de la lumière pour expliquer la transmission de l'image à l'intérieur du corps. Il a fini par comprendre que l'inversion de l'image rétinienne ne soulève pas de problème perceptif réel[5]. Mais si elle se forme comme il le pense au fond de l'œil, donc " autour du débouché du nerf optique " (pour lequel il conserve la disposition anatomique de ses devanciers), elle ne peut plus s'y engouffrer intacte ; de plus, s'il s'agit d'une image réelle, d'une *pictura*, comment du lumineux et du coloré pourrait-il se propager dans le creux d'un nerf

5. *Ad Vitellionem...*, V, 4, *op. cit.*, 185.

qui n'est pas droit, et qui est si étroit qu'il reste nécessairement obscur[6] ? De là son appel aux " physiciens ", c'est à dire aux médecins et aux naturalistes, et l'alternative qu'il propose. Ou bien les " esprits " du nerf montent de l'œil au cerveau, mais que devient l'image en cours de route ? Ou bien ils descendent du cerveau à la rétine pour l'examiner, mais comment l'information remonte-t-elle aux instances qui la jugent, et qui sont dans le cerveau ? La réalité de l'image et sa localisation ont pour sa transmission nerveuse fait perdre toute pertinence à une analogie tirée de l'optique géométrique.

DESCARTES TRANCHE LE NŒUD GORDIEN

Descartes ne répondit pas à la question posée par Kepler. Il transforma la donne dans sa *Dioptrique* (1637) de telle sorte qu'il n'ait plus à y répondre. Et ce faisant, il restructura complètement la problématique de la vision.

Son coup de force fut de renoncer à la notion d'image, du moins d'une image conservant dans ses pérégrinations internes sa ressemblance avec l'objet : " Il faut prendre garde à ne pas supposer que pour sentir, l'âme ait besoin de contempler quelques images qui soient envoyées par les objets jusques au cerveau, ainsi que font communément nos Philosophes : ou du moins, il nous faut concevoir la nature de ces images tout autrement qu'ils ne font. Car, d'autant qu'ils ne considèrent en elles autre chose, sinon qu'elles doivent avoir de la ressemblance avec les objets qu'elles représentent, il leur est impossible de nous montrer comment elles peuvent être formées par ces objets, et reçues par les organes des sens extérieurs, et transmises par les nerfs jusques au cerveau "[7]. Pour lui, le perçu peut être significatif sans pour autant être ressemblant : " Nous devons considérer qu'il y a plusieurs autres choses que des images qui peuvent exciter notre pensée ; comme, par exemple, les signes et les paroles, qui ne ressemblent en aucune façon aux choses qu'elles signifient "[8]. Dès lors, il importe de changer la problématique de la vision, car il faut " que nous remarquions qu'il est seulement question de savoir comment [les images qui se forment dans notre cerveau] peuvent donner moyen à l'âme de sentir toutes les diverses qualités des objets auxquels elles se rapportent, et non point comment elles ont en soi leur ressemblance "[9]. Ce qui compte est de distinguer entre elles les propriétés des objets externes, non de les connaître tels qu'ils sont.

L'exemple de la couleur peut illustrer cette nouveauté. Pour Kepler encore, " les esprits [du nerf] pâtissent des couleurs et des lumières, et cette passion

---

6. *Ibid.*, V, 2, 152 : " Il est tout à fait certain qu'aucune image optique ne parvient jusque là ".

7. R. Descartes, *Dioptrique,* Discours IV, *Œuvres complètes,* in Ch. Adam et P. Tannery (eds), (ci dessous A.T.), Paris, Vrin, t. VI, 112 (souligné par nous).

8. *Ibid.*, 112.

9. *Ibid.*, 113.

est, pour ainsi dire, une sorte de coloration et d'illumination "[10]. Il n'en va plus ainsi pour Descartes : entre l'impression de rouge ou de jaune et sa cause physique, il n'y a pas de communauté de nature. La cause physique est une certaine rotation sur elles-mêmes des petites boules du second élément (dont la pression transmet le mouvement de la source lumineuse), rotation qui tire de manière spécifique les filaments nerveux qui s'épanouissent dans la rétine ; cette traction qui s'exerce jusqu'au cerveau y provoque une modification correspondante d'ordre spatial ; et c'est cette modification qui " donne occasion " à l'âme de sentir du rouge ou du jaune, selon un rapport instauré par la nature. Peu importe que la théorie cartésienne de la lumière ait fait partie de ce qu'on qualifiera après lui de roman de physique. Pour la première fois la couleur devient propriété de la lumière et non des choses illuminées, la transmission nerveuse est tenue pour spécifique et distincte de l'excitant physique qui la provoque, et l'impression proprement mentale est clairement différenciée de la modification cérébrale. Descartes explicite fort bien cette redistribution : " On sait déjà assez que c'est l'âme qui sent, et non le corps [...] Et on sait que ce n'est pas proprement en tant qu'elle est dans les membres qui servent d'organes aux sens extérieurs, qu'elle sent, mais en tant qu'elle est dans le cerveau, où elle exerce cette faculté qu'ils appellent le sens commun [...] Enfin on sait que c'est par l'entremise des nerfs, que les impressions, que font les objets dans les membres extérieurs, parviennent jusqu'à l'âme dans le cerveau "[11]. Au lieu de la dichotomie traditionnelle entre l'extérieur et l'intérieur (l'image imprimée par l'agent externe sur le cristallin étant transmise telle quelle en tant qu'image sensorielle jusqu'aux instances sentantes), Descartes distingue clairement les trois phases de la perception, physique jusqu'à l'organe récepteur, nerveuse jusqu'au cerveau, mentale en tant que phénomène de conscience.

Les conséquences

Les premières conséquences concernent la physique de la lumière. Kepler suivait encore l'idée d'Alhazen et de Vitellion selon laquelle la lumière arrache en quelque sorte une couleur enfouie dans les corps : " La couleur existe réellement dans les choses mêmes, y compris quand elles ne sont pas éclairées et donc ni ne rayonnent ni ne se font voir "[12]. En distinguant l'impression de couleur de sa cause physique, Descartes accomplissait un grand pas épistémologique. Il mettait d'abord fin à une vieille confusion entre impression sensible subjective et qualité propre de l'objet. Ce qu'on discerne visuellement n'est plus ce qui est dans la chose, ni ce qu'est la chose ; l'homme n'est plus un microcosme reproduisant le macrocosme. Du côté du corps, ensuite, il rendait

10. J. Kepler. *Ad Vitellionem...*, *op. cit.*, 152.
11. R. Descartes, *op. cit.*, 109.
12. J. Kepler, *Ad Vitellionem...*, I, appendice, *op. cit.*, 41.

caduque la vieille notion aristotélicienne de sensible propre : l'impression ressentie dépend du mouvement et de la nature du nerf, non du rapport de
l'organe à ce qu'il a pour fin de sentir ; un coup reçu sur l'œil peut faire voir
des étincelles en l'absence de lumière, comme un autre sur l'oreille fait entendre un son s'il ébranle le nerf auditif[13]. Ainsi, Descartes commençait par son
mécanisme à débarrasser l'étude de la nature des " qualités " qui l'encombraient, au profit d'une optique où la diversité des couleurs n'est plus que le
signe d'une diversité de mouvement au sein même de la matière.

Les conséquences sur la théorie de la perception ne sont pas moindres. On
ne peut plus en traiter selon une métaphore judiciaire, où l'image serait appelée
devant une hiérarchie d'instances psychiques qui l'appréhendent (au niveau du
cristallin), l'unifient à son double (au niveau du chiasma), la confrontent à son
environnement sensoriel (au niveau du sens commun), la comparent et la
jugent ; car " il n'y a pas derechef d'autres yeux en notre cerveau avec lesquels
nous la pussions apercevoir ; mais plutôt, ce sont les mouvements par lesquels
elle est composée, qui, agissant directement contre notre âme, d'autant qu'elle
est unie à notre corps, sont instituées de la Nature pour nous faire avoir de tels
sentiments "[14]. Le perçu ne se donne pas comme si dès l'organe sensoriel il y
avait une conscience latente ; il n'est présent, comme le montrent les *Méditations*, qu'à un " Je pense ", un *Cogito* qu'il faut concevoir en première personne. Mais quand il perçoit, le sujet est toujours un sujet psychophysique. Au
lieu d'insister comme les perspectivistes, pour le repérage de la situation d'un
objet ou l'évaluation de sa distance, sur l'intellection et l'interprétation par
l'âme d'une image reçue, Descartes attribue un rôle majeur à l'information
inconsciente et réglée qu'elle tire de son union intime avec le corps : convergence binoculaire, accommodation, contraction de l'iris, *etc.* ; et comme ces
adaptations corporelles atteignent très vite leurs limites, il conclut que la vue
est fort sujette à l'erreur. L'analyse de cette dernière est réorganisée en profondeur, et recherche ce qui peut tromper l'âme en tant qu'elle est unie au corps :
vapeurs du cerveau chez le rêveur ou le frénétique, mauvaise disposition des
nerfs chez celui qui voit double, anomalie de l'œil ou son inadaptation à une
disposition insolite du visible[15]. La vieille psychologie des facultés fait place
à l'étude d'un sujet qui voit selon son corps.

On ne peut qu'évoquer les conséquences philosophiques d'un tel remaniement. Une très profonde transformation est en train de s'opérer dans le rapport
de l'homme au monde. Toujours et partout, entre lui et les choses, son corps
s'interpose avec ses propres lois de fonctionnement. Il perçoit non pas ce que
sont les choses, mais au mieux qu'elles existent, et si elles sont utiles ou dan

13. R. Descartes, *Dioptrique,* Discours VI, 131.
14. *Ibid.*, 130.
15. *Ibid.*, 141-147.

gereuses pour sa survie[16]. S'il veut les connaître, il lui faut les concevoir par une physique abstraite, indépendante des sensations qu'elles provoquent. Paradoxe qui sera désormais celui de la science moderne, la science du monde sensible, la physique, ne peut plus s'appuyer sur les catégories du sensible. Et le sujet percevant, comme l'attesteront ultérieurement Malebranche ou Berkeley, ne peut plus être très sûr de l'origine de ses perceptions.

Entre la conscience percevante et la chose perçue, s'instaure un nouveau rapport, qui oblige à repenser la notion de causalité. Depuis l'antiquité, la vieille idée aristotélicienne selon laquelle l'agent cherche à rendre le patient semblable à lui s'imposait dans l'explication de la perception. La saisie de la lumière et de la couleur était pensée comme illumination et coloration des instances sentantes, qu'il s'agisse des rayons visuels chez Ptolémée ou du cristallin et du nerf optique chez Roger Bacon, Vitellion, et même encore chez Kepler. Descartes par son rejet d'une nécessaire ressemblance entre le perçu et l'objet rompt radicalement avec cette tradition. La cause, dans le cas de la couleur, peut être radicalement différente de l'effet. Mieux même, quand il s'agit de l'effet sur l'âme de la modification cérébrale qu'elle induit, elle n'est ni formelle, ni finale, et ne peut être non plus matérielle, ni par conséquent efficiente : de l'étendu (le corps) ne peut agir sur de l'inétendu (l'âme), car les deux substances étant distinctes, leurs modes le sont aussi[17]. Tout au plus leurs changements concomitants sont-ils réglés selon les lois de la nature. Qu'on s'interroge sur ces dernières et Hume n'est pas loin...

Une question se pose en des termes neufs, que le dualisme cartésien rend encore plus aiguë : comment deux sphères aussi hétérogènes, celle de l'âme, inétendue et qui pense, et celle du corps, étendu et qui ne pense pas, peuvent-elles agir l'une sur l'autre ? Pour les Anciens ou les Perspectivistes, la contagion analogique rendant le sentant semblable au sensible masquait la difficulté, au moins pour ce qui concerne la vision. Chez les cartésiens la contagion n'est plus possible, car elle se heurte à un hiatus ontologique entre deux substances différentes. A l'ancienne évidence du voir s'oppose celle de Malebranche : " Il est évident que les corps ne sont point visibles par eux-mêmes, et qu'ils ne peuvent agir sur notre esprit, ni se représenter à lui. Cela n'a pas besoin de preuve, cela se découvre de simple vue "[18]. Certes, l'impasse est celle de la théorie cartésienne de la substance, mais à travers elle on assiste à l'acte de naissance du problème des rapports de la conscience au cerveau, du *mind-body problem* dont ne sont pas encore sorties nos sciences cognitives.

16. R. Descartes, *Méditations*, VI. A.T. IX-1, 69-70.

17. R. Descartes, *Principes*, I, art. 53, 60, 61. A.T. IX-2, 48-52.

18. Malebranche, *De la Recherche de la vérité*, X[e] éclaircissement in G. Lewis (éd.), Paris, Vrin, 1946, t. 3, 74.

DESCARTES AND HOBBES ON OPTICAL IMAGES

Antoni MALET

INTRODUCTION

Kepler introduced a crucial distinction between " pictures and images ". The Keplerian image of an object is the representation" directly " perceived on a mirror, through a lens, or through any kind of refractive interface. Not physical nor mathematical entities, such images were acts of vision involving the active participation of the observer's *sensorium* — which left their locations indeterminate. Kepler took images seriously, in the sense that he provided a new mathematical construction for the cathetus rule (traditionally used to determine the locus of the image) meant to improve it. Yet, since images show objects distorted and out of their true places, he also stressed their illusory, deceitful character[1]. A Keplerian picture, on the other hand, is just a set of concurrence or focal points projected on a screen. Pictures have well defined spatial locations (moving the screen back or forth blurs the picture, because the focal points are geometrically determined), do not deceive us about the object pictured, and are invisible in themselves (they are not perceived directly but need a screen to materialise). In Kepler's thought pictures are not a special kind of images because pictures and images are different kinds of things. Eventually, in the last third of the century the English school of geometrical optics tried to collapse the two notions by giving images the geometrical scaffolding of pictures. The attempt was a failure but meanwhile James Gregorie, Isaac Barrow and Newton had laid the foundations of modern geometrical optics[2]. This paper aims to survey the two main episodes in the transformation of optical

1. A. Malet, " Keplerian Illusions : Geometrical Pictures versus Optical Images in Kepler's Visual Theory ", *Studies in History and Philosophy of Science,* 21 (1990), 1-40 ; G. Simon, " Behind the Mirror ", *Graduate Faculty Philosophy Journal,* 12 (1987), 311-350, 337-342.

2. The crucial works to keep in mind are J. Gregorie's *Optica promota* (1663), I. Barrow's *Optical lectures* (1669), and Newton's Cambridge optical lectures of 1670-72, which were not published until much later.

images from Kepler's views in his 1604 *Ad Vitellionem paralipomena* to the English contributions of the 1660s.

DESCARTES

Descartes developed Kepler's optics away from geometrical optics — in the sense that he claimed the task of geometrically dealing with Keplerian images to be impossible. He preserved Kepler's distinction between pictures and images, only to dismiss Keplerian images as irrelevant. In tune with the sceptical tradition that shaped his early thought, Descartes classed images among the deceptions of the senses and suggested they were an obsolete notion — to dismiss along with *final cause,* or *materia prima,* we may add. Yet he introduced a terminological change that did cause confusion in the historiography, for while dropping Keplerian images Descartes gave *pictures* the name *image.*

Descartes' theory of vision as well as his explanation of the telescope rest directly on the retinal picture, which he consistently termed *image* (in French). To avoid confusion, I shall keep the word " picture " to refer to Descartes' *images*. To be sure, I know of no instance in *La Dioptrique* where *image* designates images perceived on lenses and mirrors, that is, Keplerian images. When Descartes mentioned direct perceptions of objects, and represented them in his diagrams of mirrors and lenses by dotted lines, he used no specific word. *La Dioptrique* contains no articulate discussion of Keplerian images. The only reference to them is found in the few pages devoted to the topic, " The reasons why it sometimes happens that [vision] deceives us "[3]. There Descartes perfunctorily discussed, among other things, why we see two images of an object when one of our eyes is forced by a finger to look into another direction, or why things appear yellow-colored when looking through a yellow glass. In this context Descartes mentioned that through lenses objects appear out of their place and devoid of their true dimensions. In fact he exercised his sarcasm against the ancients who " would pretend to have determined the place of the images in concave and convex mirrors "[4]. In 1632, when Mersenne asked him what mathematical curve would give the shape of a refracted straight line, Descartes had said as much : " there is no fixed place for the image on reflected or refracted rays "[5].

Interestingly, in the same context where Descartes dealt with visual illusions and Keplerian images he addressed the problem of distance perception. The two questions were already connected in Kepler's *Paralipomena* — images

3. *Ie vous veux faire encore icy considérer les raisons pourquoy il arrive quelquefois qu'elle [la vision] nous trompe, La Dioptrique,* in C. Adam, P. Tannery (eds), *Œuvres de Descartes*, 13 vol., Paris, Vrin-CNRS, 1982-1991, VI, 141 (all the references are made to this edition). The subject is dealt with on p. 141-147.

4. *Ibid.*

5. *Œuvres de Descartes*, I, 255-256.

were located by visually evaluating their distances to the eye. The two questions were to remain connected through the end of the century in the geometrical optics works mentioned above. Sceptical about the power of sense perception, Descartes held our eyes to be hardly reliable at measuring distances. According to him the eye's shape hardly changes by accommodation, except when the distances involved are small. Moreover, the optical triangle based on the two eyes is not very useful for distances beyond one or two hundred feet. To conclude the few pages dealing with visual deceptions Descartes pointed out that distances cannot be accurately evaluated by means of size, shape, color, or light, because perspective drawings show us quite well how easy it is to be deceived by these features[6]. He provided a more philosophical discussion of the limitations of sense perception in the *Méditations de philosophie première*[7]. It is customary to credit Descartes with the notion that our mind uses some " natural geometry " to locate objects in space by calculating their distances. In fact there is nothing peculiarly Cartesian in the " natural geometry " of visual perception. It is true that Descartes introduced such a notion linked to the triangle determined by the observer's two eyes and the object, but just to conclude that the observer cannot accurately determine visual distances. As we shall see now, what turns out to be peculiarly Cartesian, is playing " natural geometry " down as a mechanism that provides accurate visual information.

Figure 1

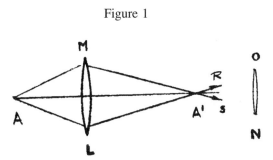

" Natural geometry " was independently introduced in optics in 1663 by James Gregorie, who made it the basis of geometrical optical images. Following Kepler's hints, Gregorie assumed we have the ability to evaluate distances visually by means of the triangle based in the observer's two eyes and by ocular accommodation (he calls it " application ")[8]. Yet already in 1651 in Paris, under the name of Mersenne, Roberval had introduced a similar notion of opti-

---

6. *Œuvres de Descartes*, VI, 144.

7. See for instance, *Œuvres de Descartes*, VII, 83 ff.

8. See A. Malet, " Studies on James Gregorie (1638-1675) ", Ph.D. Diss., Princeton University, 1989, 115-121.

cal image, albeit his was not as elaborate as Gregorie's[9]. Recognizing that the shape of the eye changes as it focuses on points at different distances, Roberval assumed that par *une longue habitude* our cognitive faculty learns there by to evaluate distances[10]. The place of the image is therefore the point or points from which the rays reaching to the eye proceed : *car ce point sera le lieu apparant de l'image exterieur du point de l'obiet dont il s'agist... car en suite de ce qui a este dit, il faudra que pour voir cette image, l'œil et [the retina] se disposent comme pour regarder un obiet qui seroit en ce mesme lieu de rencontre*[11].

In other words (see Figure 1), to the eye NO the image of point A through the lens ML is A' because the pencil of rays from A gathers in A', to flow from there towards the eye. Since the eye " applicates itself " to receive them clearly, the soul perceives A as if it were A'. Roberval was a prestigious, well-connected, first-rank mathematician, and yet his contribution went unnoticed. On the other hand, the same idea found a congenial climate in Restoration England, and geometrical optics flourished there from the 1660s to 1709, when it received a deadly blow from Berkeley's *New Theory of Vision*. The basic idea encapsulated in the optical images of Roberval, Gregorie, *et al.* obviously runs head-on against Descartes' main thesis that *nous ne pouvons d'ordinaire juger avec assûrance de la distance des objects*. The quotation comes from Malebranche, who developed at length Descartes' thought and did much to ensure that Cartesianism in one or other guise was a dominant intellectual force in most Continental circles through the second half of the seventeenth century and beyond[12]. Descartes and Malebranche stressed the empirical fact that we cannot discern whether an object moves away from, or towards us, nor can we sensibly guess a distance when the distances involved are not rather short[13]. Interestingly we find the same argument in Continental authors, like Huygens, that could not accept the new notion of optical image. The new notion of optical image, therefore, was differently received in Britain and on the Continent because in Britain it benefited from a pervasive empiricist attitude towards knowledge[14].

9. M. Mersenne, *L'Optique et la catoptrique*, Paris, 1651, 112-114. On Roberval's optics and his contribution to Mersenne's book, see R. Lenoble, " Roberval 'editeur' de Mersenne et du P. Niceron ", *Revue d'histoire des sciences,* 10 (1957), 235-254 ; and L. Auger, *Un savant méconnu : Gilles Personne de Roberval (1602-1675),* Paris, 1962, 55 ff.

10. M. Mersenne, *L'Optique et la catoptrique*, 112.

11. *Ibid.,* 114.

12. N. Malebranche, *Recherche de la vérité,* , 3 vol., Paris, Vrin, 1962-64, 108.

13. *Ibid.,* 111-113, 119.

14. G.A.J. Rogers, " The Empiricism of Locke and Newton ", in S.C. Brown (ed.), *Philosophers of the Enlightenment,* Atlantic Highlands, 1979, 1-30 ; " Locke's Essay and Newton's Principia ", *Journal of the History of Ideas,* 39 (1978), 217-232 ; " Locke, Newton, and the Cambridge Platonists ", *Journal of the History of Ideas,* 40 (1979), 191-205 ; R.W.F. Kroll, " The question of Locke's relation to Gassendi ", *Journal of the History of Ideas,* 45 (1984), 339-361 ; J.W. Yolton, *John Locke and the Way of Ideas,* Oxford, 1956, chapter 11, 26-48.

Let us stress finally that distance perception, even the limited ability to perceive it granted by Descartes and his followers, came to be linked to Christian apology. In Descartes " natural " evaluations of distance stand for soul-performed computations. In Malebranche they stand for God-performed computations, that is, computations *que nous pourrions les former nous-mêmes, si nous sçavions divinement l'optique & la géométrie, [et] tout ce qui se passe actuellement dans nos yeux & dans nôtre cerveau.* According to Malebranche, everybody knows that the soul has not got such information (*connaissances*) nor the capability (*puissance*) to do so, and to call " natural " certain mechanisms involved in sense perception is just a way to mark it out : *j'ai appellé naturels ces sortes de jugemens [sur la distance, grandeur, &c. des objects] pour marquer qu'ils se font en nous, sans nous, & même malgré nous*[15]. Even in Descartes' more restrained version the visual apprehension of distances could be an argument against materialism and atheism. At least in England the theological dimension of views on distance perception was fully recognized. The latitudinarian thinker, Edward Stillingfleet, used the mind's apprehension of things not conveyed to it by the senses, and lie mentioned distances explicitly, as an argument against Hobbes' materialistic philosophy[16].

HOBBES

According to Hobbes, perception is essentially mechanical and passive, which was consistent with his thesis that *all* effects in nature are caused by a single universal primary uncaused principle, motion[17]. As we shall see now, his mechanistic understanding of vision gave an original twist to Descartes'. He criticised Descartes' notion that an immaterial soul was the ultimate seat of perception on the basis that only body can receive motion[18]. He acknowledged

15. N. Malebranche, *Recherche de la vérité,* 119-120.

16. S.I. Mintz, *The Hunting of Leviathan. Seventeenth-Century Reactions to the Materialism and Moral Philosophy of Thomas Hobbes,* Cambridge, Cambridge University Press, 1970, 152, without reference.

17. F. Brandt, *Thomas Hobbes' Mechanical Conception of Nature,* Copenhagen and London, Levin & Munksgaard and Hachette, 1928, is still the most comprehensive account of Hobbes' natural philosophy. The main primary sources on Hobbes' optics are E.C. Stroud, " Thomas Hobbes' *A Minute or First Draught of the Optiques* : A Critical Edition ", Ph.D. Diss., The University of Wisconsin Madison, 1983 and *De homine,* in T. Hobbes, *Opera philosophica quae latine scripsit,* 5 vol., W. Molesworth ed., London, 1839, 11, 1-132 (without figures). There is a French translation which reproduces the figures, but must be used with extreme caution : *De homine. Traité de l'homme,* P.-M. Maurin trans. and ed., Paris, Blanchard, 1974. On Hobbes' geometrical optics, see A. Malet, " Hobbes on optical images : Mathematics and metaphysics in Hobbes' geometrical optics ", forthcoming.

18. Criticism of Descartes' theory of vision is absent from the " First Draught " and *De homine,* but is explicit in the earlier " Tractatus opticus II " ; see J. Bernhardt, " La polémique de Hobbes contre la *Dioptrique* de Descartes dans le *Tractatus opticus II* (1644) ", *Revue Internationale de Philosophie,* 33 (1979), 432-442 ; F. Alessio, " Thomas Hobbes : Tractatus Opticus (Harley Mss 6796, ff. 193-266) ", *Rivista critica di storia della filosofia,* 18 (1963), 147-228, 204, 207-208, 222-223. See also Hobbes, *Correspondence,* 60, n. 4.

the formation on the back of the eye of a Keplerian picture, but took it to be just a by-product of refraction in the ocular humours that did *not* mediate vision. Hobbes held vision to be accomplished through the optical axis, the perpendicular going from the point perceived to the eye and through its centre. The motion we call light reaches the retina through this line and generates a resistance or contrary motion in the same direction but outwards. It is this motion stirred in our spirits that we identify as the image and locate in the direction of the optical axis. When the thing perceived has a surface, the optical axis sweeps quickly across the surface, and its image is obtained by putting together in the memory all the images produced. Light, colour, images of objects in lenses and mirrors, the objects as we see them, or the memories we have of them are but " fancies " or " phantasms " in our mind. By stressing their phantasmic character, Hobbes excluded " Sense and Memory of things " from the legitimate sources of knowledge more radically that Descartes did.

The same is true of Hobbes' views on distance perception, for it was his thesis that objects systematically appear smaller and closer to the observer than they are. His mechanistic understanding of vision led him to assume the existence of a sensory threshold below which mechanical stimulation produces no vision. Now, objects contain small parts that fall below the sensory threshold (that small parts do " disappear " is deduced from the observation that we do not see the *whole* sun, for instance, but a reduced image of it)[19]. Yet, since the object is perceived, there must be a shorter distance at which the actions of the small parts added together will produce an image[20]. It may seem paradoxical that Hobbes compares physical distances with the " distances " at which mental constructs appear. To compound it, Hobbes geometrically " determines " the apparent places of things.

---

19. *In qualibet linea sumi potest pars aliqua adeo exigua, ut per se nullam omnino faciat sui visionem. Est enim ad sensionem faciendam in omni objecti ab oculo remotione certa quaedam et finita vis necessario requisita, ita ut minor ea, etsi moveat organa visionis, non tamen sensibiliter id faciat. Sed quodlibet objectum divisibile est in partes perpetuo divisibiles. Quare, infinities repetita divisione, pervenitur tandem ad partem tam exiguam ut vis ejus sentiri non possit* (De homine, 19).

20. *Intelligatur A pars objecti adeo exigua, ut a distantia AE, sola existens, non possit videri ; similiter sumatur alia pars H, quae sit ipsi A aequalis,... ; movet tamen oculum utraque, etiam sola, nempe pars A per rectam AE, et pars H per rectam HE,... Quoniam autem conjunctis ipsarum partiumque intermediarum viribus fit visio totius AH, id est, imago ejus aliqua, imago illa apparebit ex ea parte ubi vires conjunguntur, id est, ex ea parte ubi AE, HE convergunt, id quod est citra objectum verum AB alicubi, puta in FI ; ...Ob eandem rationem videbitur BK in GL, et totum objectum AB in FG* (De homine, 20).

Figure 2

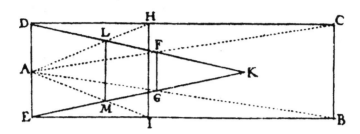

In Figure 2, A is the eye, BC an object, FG its image, and K its vanishing point. Hobbes argues that the position of K determines that of FG, and conversely, and that when any of the two is known, the image of BC at any other distance is determined as well. Let DE, centred in A, be parallel and equal in length to BC. Hobbes' construction assumes that, were the object placed in the eye, like DE, and then removed towards BC, it would appear ever smaller until, when reaching BC, it would appear at FG. Removed further backwards it would end up appearing as the point K, and then would disappear. Now, to determine the image of BC when it lies at HI, Hobbes argues that H must appear on its own visual line HA — because it is the line that follows the motion coming from H and acting on the eye. Now the image of H must be on both DK and AH, and is therefore L. Likewise, I must appear on AI and its image is M, on the intersection of AI and EK. Therefore, the image of HI will be LM[21]. Notice that the geometrical construction based on Figure 2 can never locate images in physical space. One needs to know first the location of at least one image (or, alternatively, that of the vanishing point, K) to determine the location of other images geometrically. But Hobbes explicitly claimed that the location of images cannot be determined because images are no real things[22]. What is then the purpose of the foregoing geometry ? What Figure 2 does determine is the relative position of images. It shows that images do not appear in unpredictable ways, and that the geometry of the real world somehow gets in and rules among images too. Hobbes' geometry of images would therefore fulfil the non trivial function of showing that the world of appearances has a geometrical consistency of its own. Hobbes' results on the images produced by mirrors, lenses and optical instruments also establish geometrical relationships between visible objects and their reflected or refracted images.

Hobbes kept the distinction between picture and image but turned Descartes' views upside-down. Pictures were mathematical entities with no role in

21. *De homine*, 21-22.

22. *Notandum autem est, quod etsi distantia objecti ab oculo vera mensuris determinari potest, locus tamen apparens, id est, distantia imaginis ejus in visione directa mensurari non potest, quoniam imago res est mere phantastica* (*De homine*, 22).

Hobbes' theory of vision nor in his account of telescopes, but he believed that his geometrical theory of optical images was powerful enough to " explain " well known optical effects and the working of optical instruments. Interestingly, after writing an extensive treatise on optics, the only part he gave to the printer was his geometrical theory of optical images. Hobbes was innovative in casting subjective images in mathematical terms, thereby making them able to support, so to speak, a rational, consistent, intersubjective account of visual phenomena[23]. In this sense Hobbes was adding a new dimension to Descartes' dioptrics, where images have no place. By endowing subjective sensory impressions with objective geometrical features, Hobbes suggested the virtuality of a mathematical science of visual appearances.

Now, Hobbes' geometrical optics obviously embodied crucial tenets of his materialistic philosophy. For any mathematically minded member of the English establishment, Hobbes' geometrical optics must have been a powerful stimulus to offer a different kind of optical science. The challenge was met by the geometrical optics of Barrow and Newton, which represents the obverse of Hobbes' theory of images. Naively empiricist, the new geometrical optics uncritically assumed the visual ability to apprehend accurately the material world.

---

23. " First Draught ", 385.

PART TWO

ASTRONOMY

INTRODUCTION

Suzanne DÉBARBAT

Qu'aurait été l'évolution de l'astronomie sans le développement de l'optique ? Quant à ce dernier domaine il a trouvé un terrain de prédilection dans les applications qu'en ont fait les astronomes.

Mécanique céleste s'appliquant principalement au système solaire, marginalisation d'une tradition astrologique pourtant rémanente, création de modèles, analyse de données..., tout concourt à former les esprits — au siècle des Lumières notamment — à la compréhension du Monde que viennent renforcer les découvertes de l'astrométrie et de l'astrophysique.

L'unité du volume, confortée par la variété des sujets traités, doit permettre au lecteur qui n'a pu se rendre à Liège en 1997, d'apprécier l'infuence réciproque précoce des deux domaines de la recherche qu'il couvre.

HIPPARCHAN PRECESSION-MATH SPHERICAL TRIG RELICS. MANITIUS' DISCERNMENT. ANCIENT CALCULUS ?

Dennis RAWLINS

In *Alm* 7.3, Ptolemy chooses six stars' data, with which to demonstrate precession by examining precessional motion in declination δ, between the star-catalog epochs of Hipparchos (-126.278) and himself (+137.547).

For each star, Ptolemy provides the east-west coordinate in crude zodiacal terms. E.g., " near middle of Taurus ", which of course means about 45°, since each sign is 30° wide, and Taurus is the 2nd sign of the zodiac. But if these positions are λ, then some are grossly erroneous. (For 1/2 the sample, the data are even in the wrong zodiacal sign. For star #5, Libra is cited when actual λ is in Leo, two signs away).

So it has been widely believed that Ptolemy committed several huge, incoherent scribal errors[1]. But there is a potential solution : the " zodiacal " coordinates are not λ (ecliptical) [a point effectively anticipated by K. Manitius] but are instead right ascensions α (equatorial). If so, then all the east-west values' errors shrink to just a few degrees (appropriate to the data's rough precision) ; and the *Alm*-reported zodiac sign matches reality, for each of Ptolemy's six stars. Thus, Ptolemy was not guilty of the blunders suspected of him. (Using zodiac signs for α seems odd practice today, but Hipparchos is believed to have engaged in it[2]). In the accompanying Tables 1-3, coordinates are based on the Ancient Star Catalog *(Alm* 7.5-8.1), precessed for past epochs at *(Alm* 7.4) 1° per 100 Egyptian yrs (365d each) : δ & β are expressed in degrees, while λ and α (and μ — see below) are expressed in 30° zodiacal units. (So, e.g., η UMa's $\lambda = 5.91$ for -129 refers to 91% of the way through

1. See the astonished puzzlements of G.Toomer, p. 333 fn. 62, 63, in his 1984 edition of the *Alm.*

2. See O. Neugebauer *History of Ancient Mathematical Astronomy,* 1975, 278 ; G. Toomer, *Dictionary of Scientific Biography,* vol. 15, 217.

the 5th constellation, Leo — or 4.91.30° = 147°, while η UMa's α = 7.14 similarly means 14% through Libra, or 6.14.30° = 184°). Note :

[a] The most convenient formula for declination-precession δ is the simple expression :

$$\delta = p \sin \varepsilon \cos \alpha = D \cos \alpha \qquad (1)$$

(Where D is the maximum δ precession for the time interval. See the Tables' headers). Here, for Ptolemy, obliquity ε = 23°51'20", longitudinal precession p = 1°/100[y]. (Ptolemy believed ε and p to be constant). So, in the 2/3 centuries since Hipparchos : net declination-precession $\Delta\delta$ = (2°2/3) sin 23°51'20" cos α = D cos α = 65' cos α. Indeed, for stars near the equinoctial colure (α = 0° or 180°), *Alm* 7.3 says that $|\Delta\delta|$ = 65' or 66'.

[b] In no other part of the *Alm* are celestial objects' α given.

[c] Modern non-differential calculation of precession by straightforward means (as against Manitius' neater method, cited below) would use λ (before&after) and latitude β (taken as constant) :

$$\Delta\delta = \delta_2 - \delta_1 = \arcsin[\sin \varepsilon \cos \beta \sin \lambda_2 + \cos \varepsilon \sin \beta] - \arcsin[\sin \varepsilon \cos \beta \sin \lambda_1 + \cos \varepsilon \sin \beta](2)$$

where subscripts 1&2 signify before&after, respectively, and $\lambda_2 - \lambda_1$ = 2°2/3. Also :

[1] Ptolemy treats δ as a function of only one variable, the east-west coordinate.

[2] His precession-discussion is so differential that he gives (per star) only one value for that coordinate. (Not a before&after pair).

The foregoing points are consistent with the speculation that ancient astronomers (e.g., Hipparchos) & mathematicians (e.g., Ptolemy) had access to eq. 1, an equation of differential spherical trigonometry. However (though I initially took this as quite convincing indication of calculus' antiquity), consistency does not prove truth. It must be noted that the conservative alternate explanation by K. Manitius (1912-1913, pp. 20 f) ingeniously produces the same results without calculus, by using the longitude μ (polar longitude) & declination v of the ecliptic point whose α is the star's[3]. Manitius' method compares v for two μ differing by 2°2/3. This procedure is less laborious than eq. 2 (though more so than eq. 1). In its favor : [A] Text speaks of shift along ecliptic. [B] The μ for 5 of the 6 stars are (as Manitius was first to reveal) given in Hipparchos' *Commentary*. Contrarily : [a] Point [2], above. [b] For star #1, μ = 32° = Tau 02° for Ptolemy; but the zodiacal sign attested in *Alm* 7.3 is Ari. However, star #1 is no problem if we accept Hipparchos as author [below], since its μ was Ari 29° at his epoch (2°2/3 of precession prior to Ptolemy's epoch).

---

3. I thank Hugh Thurston for triggering my appreciation of the Manitius 1912- 1913 *Alm* edition's perceptiveness regarding *Alm* 7.3, as against the now-over-dominant but also valuable Toomer edition's.

Several symptoms point to Hipparchos as originator of the east-west coords (α and μ are both equatorial-frame-based & never differ by more than $2°.6$ — not enough for sure textual discrimination here). Which indicates that the *Alm* 7.3 precessional analysis is his.

[i] Hipparchos evidently expressed both α & μ in zodiacal terms.

[ii] The $\Delta\delta$ data fit Hipparchos' epoch better than Ptolemy's. (Compare residuals in Tables 1-3.)

[iii] Historically, Hipparchos is strongly associated with precession.

So I suggest that the *Alm* 7.3 mathematical discussion and the stellar α or μ may have been taken over from Hipparchos' famous publication on precession (cited at *Alm* 7.1), long believed to have been (except for δ data) utterly lost.

A more thorough version of the foregoing explorations will appear in an imminent number of the scientific-history journal *DIO*.

Table 1   Epoch = -194 $\Delta\lambda$ = 1.667 ε = 23.9167   D = $\Delta\lambda.\sin\varepsilon$ = 40.54

#	Star	Alm 7.3 Sign	Cat #	Cat λ	Cat β	Cat μ	Cat ν	Cat α	Cat δ	TH$\Delta\delta$	HP$\Delta\delta$	Pt$\Delta\delta$	D.cos α	Resid	w
1	ηTau	1: near end	410-1	1.99	3.75	1.94	11.05	1.87	15.08	40	41	41	36.40	4.23	0
2	αAur	2: near mid	222	2.72	22.50	2.47	16.39	2.38	40.07	24	29	30	30.34	-0.34	
3	γOri	2: 2nd sect	736	2.69	-17.50	2.84	19.46	2.76	1.39	36	26	25	24.50	0.50	
4	αVir	6: 1st sect	510	6.78	-2.00	6.75	3.04	6.77	0.86	-48	-41	-41	-40.25	-1.00	
5	ηUMa	7: 1st sect	035	5.88	54.00	7.13	-1.63	7.12	60.58	-45	-41	-41	-40.46	-0.17	
6	αBoo	7: 1st sect	110	6.79	31.50	7.31	-3.73	7.28	31.04	-30	-44	-44	-40.10	-3.65	0

Sum of Weighted Residual-Squares = 1.40

Table 2   Epoch = -128.943 $\Delta\lambda$ = 1.667 ε = 23.9167   D = $\Delta\lambda.\sin\varepsilon$ = 40.54

#	Star	Alm 7.3 Sign	Cat #	Cat λ	Cat β	Cat μ	Cat ν	Cat α	Cat δ	TH$\Delta\delta$	HP$\Delta\delta$	Pt$\Delta\delta$	D.cos α	Resid	w
1	ηTau	1: near end	410-1	2.01	3.75	1.96	11.29	1.89	15.31	40	41	41	36.20	4.42	0
2	αAur	2: near mid	222	2.74	22.50	2.50	16.61	2.41	40.26	24	29	30	29.99	0.01	
3	γOri	2: 2nd sect	736	2.71	-17.50	2.86	19.60	2.78	1.55	36	26	25	24.16	0.84	
4	αVir	6: 1st sect	510	6.80	-2.00	6.77	2.78	6.79	0.60	-48	-41	-41	-40.30	-0.95	
5	ηUMa	7: 1st sect	035	5.91	54.00	7.15	-1.88	7.14	60.31	-45	-41	-41	-40.43	-0.20	
6	αBoo	7: 1st sect	110	6.81	31.50	7.33	-3.98	7.30	30.78	-30	-44	-44	-40.04	-3.71	0

Sum of Weighted Residual-Squares = 1.64

Table 3   Epoch = +137.547   Δλ = 2.667   ε = 23.8556   D = Δλ.sin ε = 64.71

#	Star	Alm 7.3 Sign	Cat #	Cat λ	Cat β	Cat μ	Cat ν	Cat α	Cat δ	THΔδ	HPΔδ	PtΔδ	D.cos α	Resid	w
1	ηTau	1: near end	410-1	2.10	3.75	2.05	12.23	1.98	16.24	64	65	65	56.40	8.60	0
2	αAur	2: near mid	222	2.83	22.50	2.60	17.48	2.51	41.00	38	46	48	45.44	2.56	
3	γOri	2: 2nd sect	736	2.80	-17.50	2.94	20.11	2.86	2.12	58	42	40	36.30	3.70	
4	αVir	6: 1st sect	510	6.89	-2.00	6.86	1.70	6.87	-0.48	-77	-66	-66	-64.56	-1.44	
5	ηUMa	7: 1st sect	035	5.99	54	7.23	-2.82	7.21	59.24	-72	-65	-65	-64.31	-0.69	
6	αBoo	7: 1st sect	110	6.90	31.50	7.41	-4.96	7.38	29.73	-48	-70	-70	-63.45	-6.55	0

Sum of Weighted Residual-Squares = 22.82

ON THE MEASUREMENT OF CELESTIAL LONGITUDE IN ANTIQUITY

Nicholas KOLLERSTROM

" For Babylonian and early Greek astronomy the beginning points of the signs are rigidly connected, not with the equinoxes, but with the fixed stars "

van der Waerden[1]

INTRODUCTION

How, in the Hellenistic world of late antiquity, was celestial longitude measured ? For over a millennium planetary longitudes were measured with respect to the Vernal Point — as Claudius Ptolemy advocated, writing in the second century AD. However, the zodiac framework of antiquity was inherited from the Babylonians, who used a sidereal reference, with certain fixed stars defining its position. Otto Neugebauer's *Greek Horoscopes* argued that these Hellenistic charts were using a sidereal reference, but was vague about what this was.

Almost nothing is written about this subject in modern histories of science or astronomy. Evidence is here presented supporting Neugebauer's view, based on computer reconstruction of planetary longitudes as given in the 1st-5th centuries AD, and some earlier, showing that a sidereal zodiacal reference continued to be used by Greek sources, comparable to that used by the Babylonians ; and supporting a claim recently made by John Britton, concerning the star β-Gemini as defining zero degrees of Cancer[2].

In the fifth century BC, a twelvefold division of the ecliptic into equal intervals appeared, which by the fourth century was clearly subdivided into thirty degrees per sign, and stellar longitudes were thereby defined ; and then, in the third century, planetary degree-longitudes appeared, for distinct moments in time. This not unremarkable evolution remains a largely ignored topic, with lit-

1. B.L. van der Waerden, *Science Awakening II*, 1974, 222.
2. J. Britton, Ch. Walker, " Astronomy and Astrology in Mesopotamia ", in Ch. Walker (ed.), *Astronomy before the Telescope*, 1996, 49.

tle having appeared since the researches of Otto Neugebauer and Van der Waerden earlier in the century.

Two books entitled *Early Astronomy* have appeared recently. One of them, by O'Neill, gave no hint that any issue over zodiac position existed or required discussion[3] ; while the other, by Thurston, posed the question : " Where precisely on the ecliptic were the signs placed ? " a propos of Babylonian astronomy[4], but then gave an answer in terms of the equinoctial points: it averred that the signs had been fixed with regard to so many degrees from the solstices and equinoxes. This, however, was far from being the case, as it was in the nature of the Babylonian zodiac that it remained fixed against the stars. The solstices and equinoxes moved round against it, at the rate of precession, viz. one degree per 72 years.

Noel Swerdlow's *The Babylonian Theory of the Planets* has reviewed the manner in which Babylonian scribes computed planetary longitudes, while avoiding any statement as regards what reference was used for such. Its graphs depict planetary longitudes through the zodiac signs, and over thirty tables show planetary ingresses into zodiac signs, but nowhere is a reference indicated whereby these divisions were fixed. Swerdlow even avers that " the division of zodiacal signs into 30 ūs does not indicate a coordinate of longitude ", and that " we shall, for lack of any alternative term, refer to locations and distances in ūs through the zodiac as longitude, and measure them in degrees, but without implying that they are measured along a circular co-ordinate "[5]. This book's strange claim, that the Scribes of Chaldea were unable to measure celestial longitude, has been described by a reviewer as " almost certainly untrue "[6]. The " ūs " was a time-measurement equal to four minutes of time, or 1/360 of a day. But, to quote Morrisson and Stephenson, " since this unit [the ūs] was the interval required for the celestial sphere to turn through 1°, it is customary to translate ūs directly as degree "[7] ; just as today, one may add, the terms " minute " and " second " signify degree measure as well as time.

STELLAR LONGITUDES

In 1958, Peter Huber published the result of his investigation into the celestial longitude framework utilised in the Babylonian astronomical data, over the period -124 to - 99[8]. He expressed his findings in terms of a single reference,

3. W.M. O'Neill, *Early Astronomy from Babylonia to Copernicus,* Sidney U.P., 1986.

4. H. Thurston, *Early Astronomy,* Springer Verlag, 1994, 68.

5. N. Swerdlow, *The Babylonian Theory of the Planets*, Princeton, 1998, 35.

6. J. Britton, *Planetary Theory in Babylon,* review of N. Swerdlow, *op. cit., Journal for the History of Astronomy,* 29 (1998), 381-386.

7. L.V. Morrison, F.R. Stephenson, " Contemporary Geophysics from Babylonian Clay Tablets ", *Contemporary Physics,* 38 (1997), 13-23, 16.

8. P. Huber, " Ueber den Nullpunkt der Babylonischen Ekliptic ", *Centaurus,* 5 (1958), 192-208. NB, the year -99 is equivalent to 100 BC.

an estimate of what will here be called the sidereal zodiac position (SZP). This refers to the angular distance in celestial longitude between the historic zodiac (which was sidereal, i.e. star-defined) and the modern value computed for the epoch in question (which is tropical, ie equinox-defined). Huber estimated this parameter to be 4°28' for the epoch of -100 BC.

Huber derived that value using merely eleven star-positions, thereby expressing the mean displacement of their stellar longitudes from those of the tropical zodiac, in that year. Six of these had come from a pre-Seleucid (*ca* 400?) fragment of a star catalogue published by Sachs[9], and five more were added by Huber from his analysis of Almanacs for -122 and -110, i.e. roughly three centuries later, an analysis which also confirmed three of the longitudes found by Sachs (and disagreed with none).

The brightest zodiac star is Aldebaran the " Bull's Eye ", and in the Babylonian 30° " sign " of the Bull of Heaven, its tropical longitude was 10°34' of Taurus for the epoch of -100[10], which is 4°26' short of the sign's mid-point, ie 15°. That difference is, for that epoch, a mere few arcminutes from the mean dispacement of the Babylonian zodiac which Huber ascertained. In other words, if a sidereal zodiac were to be defined by adding on Huber's 4°28' to tropical longitudes, for 100 BC, then Aldebaran would have a longitude of 15°02' Taurus. This suggests that they took Aldebaran at 15° Taurus, or something close to it, as a prime reference, though Huber refrained from drawing such a conclusion.

No Babylonian tablets have been found specifying a procedure for locating the sign divisions. There are a couple of references from later Hellenistic sources affirming that Antares and Aldebaran were located at central 15° positions, by Kleomedes in the 4th century and Rhetorios in the 6th century[11]. Figure 1 shows the zodiac structure using this as its reference-axis, bisecting the signs of Taurus and Scorpius. The two first-magnitude zodiacal stars Antares and Aldebaran were 180° apart to within a single arcminute over the period 300 BC to 1200 AD[12] ; and that, although they may " never be seen simultaneously in the Mediterranean area ", on some days each year they could be seen together from below the Tropic of Capricorn[13].

9. A. Sachs, " A Late Babylonian Star-Catalogue ", *Jnl Cuneiform Studies*, 6 (1952), 146-150.

10. P. Huber, *op. cit.*, 205.

11. O. Neugebauer, *A History of Ancient Mathematical Astronomy*, II, Berlin 1975, 960, assumed that the positions for Aldebaran and Antares given by Kleomedes (1st century BC) were tropical, but there is no clear evidence either way. Rhetorius gives 16°20' Taurus/Scorpio for these longitudes (5th century AD), which are clearly tropical.

12. D. Rawlins, *Dio*, (Jan 1991), 62 (Rawlins Ed., an independent US history of astronomy journal ; copies in Royal Astronomical Society library, London).

13. K. Pickering, *Dio,* (April 1992), 15.

Figure 1. The Sidereal Zodiac (Source : R. Powell & P.Treadgold, *The Sidereal Zodiac*, 1979), with names of some first-magnitude stars, plus movement of Vernal Point from 5th century BC (Euctemon) to today at 25° of Pisces.

About thirty-two reference or normal stars were employed by Babylonians[14], and it may be of interest to view the positions of a few of these, converted to a sidereal zodiac with Aldebaran at 15°00' Taurus in at epoch -100 as shown in Table 1.

Table 1 : Key Star-Longitudes, Sidereal (for epoch -100)

Aldebaran	(α-Taurus)	15°00'
Antares	(α-Scorpio)	15°00'
Regulus	(α-Leo)	5°12'
Spica	(α-Virginis)	29°06'
Other stars near to sign-boundaries:		
Alhecka	(ζ-laurus)	00°02' Gemini
Pollux	(β-Gemini)	28°48'
Deneb Alg.	(δ-Capricorn)	28°41'

The stars at sign boundaries had names, eg δ-Capricorn was called " rear star of the Goat-fish "[15] which names however derived from the *constellation* rather than the *sign*.

Recently, Britton and Walker have averred that " the norm of the Babylonian sidereal zodiac appears to have been fixed so that the longitude of the bright star β-Gemini was 90° "[16], but without giving any evidence for this view. That appears to be the only post-Neugebauer statement concerning the positioning of the star-zodiac of antiquity made by science historians.

Relative to the values given in Table 1, a one-degree and twelve arcminute rotation of the zodiac would put Pollux (β-Gemini) on the Gemini/Cancer boundary. It is unclear as to whether such a shift would have been significant, as we hardly know the extent to which the reference-star longitudes were coherent. One should not assume that such positions at different portions of the ecliptic were meant to be fully compatible. Van der Waerden concluded that the Babylonians positioned the star Spica at 29° of Virgo, " with a possible deviation of 1° to either side "[17], which gives a fair idea of the accuracy involved. Four of the stars listed above probably denoted sign boundaries, with Pollux placed at 90° as the Gemini/Cancer boundary. Of the first-magnitude stars,

14. A. Sachs, *The Place of Astronomy in the Ancient World,* Ed. D.G.Kendall *et. al.,* Oxford, 1974, 46 ; the 32 reference stars are listed with ecliptic coordinates in A.Sachs and H.Hunger, *Astronomical Diaries and Related Texts from Babylonia,* 1988, 1, 17-19.

15. A. Sachs, *op. cit.,* 46.

16. J. Britton, Ch. Walker, *op. cit.,* 49.

17. B.L. van der Waerden, " History of the Zodiac ", *Archiv für Orientforschung,* 1953, 16, 216-230, 222. Spica is one of the stars included in both Sach's star catalogue fragment (ref.9) and Huber's almanac analysis (ref.8), in both cases with a longitude of 28°.

Spica and Regulus were especially useful as being on the ecliptic, having celestial latitudes of 2° and 0° respectively.

In contrast with their attention to stellar longitudes, van der Waerden emphasised how little interest the Babylonians had over solstice and equinox positions : " they did not care for exact observation of equinoxes... the spring equinoxes have errors up to 5 days "[18]. There is a profound contrast here with Greek astronomers who, once they heard about the zodiac later in the fifth century BC, defined it by reference to the Vernal Point, taking the longitude of this at around 8° of Aries. A zodiac thus defined is *tropical,* and this inquiry is not concerned with such. No Greek astronomer is on record as having measured celestial longitude by reference to fixed stars.

Hipparchus of Rhodes " reckoned the seasons from the beginning of the signs "[19], thereby conceiving the zodiac signs tropically. He may not have used ecliptic longitude for his measurements, but an oblique or right ascension, making it not altogether clear to historians in what way he discovered precession. Assuredly, however, he made a radical break from the ways of the Babylonians, who produced " ...Cuneiform tablets in which the vernal point is put at 10° or 8° of Aries, but never at 0° Aries "[20]. Earlier, the Greek or Tropical zodiac had been used for estimating time in terms of the seasonal course of the year, by astronomers Meton and Euctemon, whereas Hipparchus used it as the basis for cataloguing star-positions.

Van der Waerden emphasised how, after the transmission of the Babylonian zodiac to the Hellenistic world of late antiquity, the habit of measuring from a sidereal reference was retained, and that, once the notion of precession came to be accepted, astrologers realised that they only had to add on a certain number of degrees, depending on the year, to convert tropically-computed tables to sidereal. Referring to Theon of Alexandria, who lived in the fourth century, van der Waerden explained his conversion procedure, adding : " Their successors... living in the era of Diocletianus, still used the same rule, adapted to the new era. Obviously the purpose of the rule was, to reduce tropical longitudes " (taken eg from tables of Apollonius, Hipparchus or Ptolemy) to sidereal longitudes "[21]. Theon of Alexandria in his commentary on Ptolemy's *Handy Tables* in the fourth century recommended that certain longitudes be measured from the star Regulus[22].

Ptolemy introduced the Tropical reference system for measuring celestial longitude, at a time when the two systems were merely one degree apart. Was

18. *Ibidem,* 223.

19. O. Neugebauer, *op. cit.,* I, 278 ; *Hipparchi in Arati et Eudoxi Phenomena Commentarium,* Teubner, 1894, Ed. C. Manitius, 5-7, 48.

20. B.L. van der Waerden, *Science, op. cit.,* 126.

21. B.L. van der Waerden, " History... ", *op. cit.,* 228.

22. O. Neugebauer, *A History...,* *op. cit.,* 1002 ; H.T.I. Halma, *Commentaire de Théon d'Alexandrie sur le livre III de l'Almageste de Ptolémée,* Paris, 1822.

Ptolemy thereby reflecting an accepted practice ?[23] This question was here investigated by using the collection *Greek Horoscopes* by Neugebauer and van Hoesen[24]. To avoid misunderstanding, this title did not mean that the horoscopes were made either by Greeks or in Greece, but solely that they were written in the Greek language. They are mainly from Egypt.

A TEST

Twenty-one charts were located in this compendium which had planetary longitudes given in degrees (Table 2), one-fifth of the total, and the date of the chart reliably ascertained. These charts spanned the period from AD 40 to AD 497, almost five centuries. Time-of-day information was normally given for each chart, either the hour of day or the zodiac sign rising. It was decided to omit Mercury since, as van der Waerden points out, its longitude values were subject to large errors[25]. Excluding Mercury, at least four planetary (including Sun and Moon) longitudes were required. A couple of obviously mistaken positions were omitted; which was taken to mean, that their SZP differed from the mean of the others by more than twice the standard deviation of the group.

The charts were mainly from around Alexandria, and to convert its local time to GMT, two hours were subtracted. GMT was then converted into the Ephemeris time required for the computation by adding on ΔT[26], which was about two hours for the era of Ptolemy[27]. This ΔT factor expresses the gradual slowing down of Earth's rotation due to tidal friction. (Ephemeris Time is a measure of time derived from the uniform motions of the planets, while GMT is defined by the rotational motion of the Earth). Thus there are two steps of time-transformation required, to turn the historically-given times into those used for the computation.

Planetary longitudes in *Greek Horoscopes* were mostly given in degrees only, without arcminutes, in which case they sometimes gave the longitude as being thirty degrees, but never as zero degrees (Table 2). The first degree position (from zero to 59 arcminutes) was being designated as " one ", up to their last degree within a single zodiac sign (from 29 degrees, zero arcminutes up to 59 arcminutes) which they called " thirty ". Thus, our best estimate of their one

23. J. North, *Fontana History of Astronomy and Cosmology,* 1994, 67.

24. O. Neugebauer, H.B. van Hoesen, *Greek Horoscopes* 1959, American Philosophical Society, 1987.

25. B.L. van der Waerden, " History... ", *op. cit.,* 229 ; Ptolemy's Mercury theory generated regular errors of up to 6°, whereas for other planets his errors were only a degree or so. *Cf.* Owen Gingerich, *The Eye of Heaven,* New York, 1993, 67-68.

26. *Explanatory Supplement to the Astronomical Ephemeris,* London, 1992, p. K8-K9 gives ΔT values from 1620 onwards ; Strictly speaking, Ephemeris Time (ET) became Terrestrial Dynamical Time (TDT) after 1984 : *The Astronomical Almanac,* 1993, p. B4-B7.

27. F.R. Stephenson, " Long-Term fluctuations in the Earth's Rotation : 700 BC-AD 1990 ", *Philosophical Transactions,* Ser. A, 351 (1995), 165-202.

degree is 0.5°, etc. Half a degree was therefore subtracted from each given degree-value where no arcminute values were given.

Table 2 : Historical planetary longitudes

The planetary longitude-values, 0-30°, as given in ancient horoscopes, with the modern-equivalent Julian year and date, in degrees (d) and minutes (m), plus given zodiac signs.

	Sun	Moon	Saturn	Jupiter	Mars	Venus	Mercury
	d m	d m	d m	d m	d m	d m	d m
BABYLONIAN HOROSCOPES							
−234 June 3	12 30 ♊		6 ♋	18 ♐	24 ♋	4 ♉	
−199 June 4		15 ♋	10 ♏	26 ♏	10 ♉	5 ♊	27 ♊
−198 Oct 31			3 ♎	10 ♑	10 ♐	4 ♑	8 ♏
−87 Jan 5		5 ♉		27 ♈	20 ♌	1 ♓	26 ♐
−68 April 15	30 ♈	18 ♑	15 ♒	24 ♑	14 ♎	13 ♊	
GREEK HOROSCOPES							
40 April 5	19 ♈	15 ♓	20 ♎	6 ♒	15 ♈	5 ♈	6 ♈
46 Jan 3	11 ♈	11 30 ♑	30 ♐	19 ♋	14 30 ♒	19 ♑	30 ♐
75 July 19	29 30 ♋	12 ♓	27 8 ♐		7 23 ♍	28 13 ♋	11 25 ♌
76 Jan 24	8 ♒	1 ♒	5 ♑	1 ♒	22 ♓	12 ♓	12 ♑
81 March 31	14 6 ♈	13 ♉	5 59 ♓	6 ♋	16 3 ♏	16 4 ♓	10 ♈
110 March 15	25 8 ♓	16 53 ♊	1 25 ♓	25 18 ♐	21 ♉	8 ♒	
137 Dec 4	13 23 ♐	3 6 ♏	3 38 ♏	12 44 ♈	30 0 ♑	9 54 ♐	15 2 ♐
260 Sept 29	8 ♎	8 32 ♑	11 32 ♈	3 ♏	14	8 16 ♏	23 44 ♏
338 Dec 24		9 1 ♐	22 ♓	24 23 ♉	14 6 ♑	29 ♏	19 ♑
412 Feb 8		17 29 ♊	24 23 ♉	24 41 ♉	29 50 ♐	23 ♓	4 42 ♒
440 Sept 29	5 8 ♎	8 4 ♉	25 ♉	23 8 ♎	26 8 ♑	26 ♏	23 ♎
463 Apr 25	3 ♉	23 ♑	26 ♒	17 ♏	1 ♈	15 ♈	26 ♉
474 Oct 1		27 36 ♏	27 ♋	16 30 ♏	17 30 ♈	23 30 ♌	25 30 ♎
478 Aug 29	4 ♍	9 ♓	11 ♏	26 ♐	20 ♊	24 ♍	18 ♌
479 July 14	19 ♋	16 ♍	16 ♍	8 ♒	18 ♍	6 ♊	4 ♌
482 Mar 21	29 43 ♓	6 ♎	25 ♎	28 ♈	29 ♓	15 ♉	20 ♓
483 July 8	18 ♋	10 ♒	11 ♍	19 ♊	6 ♎	7 ♌	18 ♋
484 July 18	23 ♋	7 ♍	15 ♍	20 ♋	5 ♋	26 ♊	19 ♌
486 March 17	26 ♓	27 ♑	11 ♐	27 ♌	25 ♈	2 ♓	23 ♓
487 Sept 5	10 ♍	4 ♎	15 56 ♐	7 55 ♎	8 ♑	8 ♌	25 ♏
497 Oct 28	4 22 ♏	18 10 ♉	24 28 ♈	4 12 ♏	17 35 ♌	21 58 ♏	14 32 ♏

For each of the times at which these charts were drawn up, the longitudes of the Sun, Moon and planets excluding Mercury were ascertained using a modern astronomical program (Table 3)[28]. These were then subtracted from the corresponding values given in the charts[29], and the mean SZP value computed for each chart (Table 4). Plotting these as a function of year to which the chart belonged gave Figure 2. A best-fit regression line has been added, which gave a slope of one degree per 72.8 years, a value essentially identical with the rate of precession.

28. The program " Electric Ephemeris " (UK) Windows version 3.1 was used.

29. The 486 and 487 charts were assumed to have suffered from a " very common scribal error " which put 12 instead of 2° (Venus 486) and 18 instead of 8° (487 Mars) : advice from Anne Tihon.

This data indicates that longitude values in the collection of charts were moving systematically with respect to the tropical (sun-defined) zodiac reference, indicating that astrologers of this period were using a star-defined zodiac and not a tropical zodiac. It would appear that their practice was still, to use van der Waerden's phrase, " rigidly connected " to some stellar framework, one moving by one degree per 72 years against the tropics. This has implications for the status of the tropical framework advocated by Claudius Ptolemy (following Hipparchus) for measuring celestial longitude in both his *Almagest* and his *Tetrabiblos* : to the effect that, while this may have been used by some astronomers of the time, it was not used by astrologers for drawing up their charts.

Table 3 : Modern (computed) longitude values, 0-30 deg.

	Sun		Moon		Saturn		Jupite		Mars		Venus		Mercury	
	d	m	d	m	d	m	d	m	d	m	d	m	d	m
BABYLONIAN CHARTS														
−234 June 3	7	16			24	17	13	38	19	20	21	38		
−199 June 4			12	39	1	27	22	2	1	50	27	0	16	43
−198 Oct 31					27	22	29	25	2	16	21	40	11	43
−87 Jan 5			1	13			21	30	16	44			22	12
−68 April 15	22	37	13	20	10	12	17	49	5	59	7	47		
GREEK CHARTS (AD)														
40 April 5	13	45	12	14	17	23	28	54	8	0	28	47	25	6
46 Jan 3	12	25	8	21	21	54	22	49	18	40	12	44	26	5
75 July 19	24	5	11	59	17	48			9	30	21	38	9	18
76 Jan 24	2	42	27	37	29	57	26	33	19	14	14	3	6	55
81 March 31	9	19	12	11	0	57	0	21	13	17	7	32	11	6
110 March 15	23	38	15	55	24	42	21	30	13	4	8	18	11	52
137 Dec 4	11	55	33	10	27	19	10	18	32	18	11	43	20	11
260 Sept 29	6	27	10	35	5	51	3	37	0	13	11	11	27	36
338 Dec 24					9	51	21	37	14	57	29	6	9	51
412 Feb 8			17	12	27	36	25	45	33	1	23	47	6	48
440 Sept 29	7	48	12	45	27	30	24	58	28	15	22	19	27	20
463 Apr 25	5	31	25	58	28	21	18	16	2	41	10	44	25	59
474 Oct 1			31	5	31	5	18	11	19	36	26	43	2	28
478 Aug 29	7	12	13	19	13	59	27	3	22	5	27	59	19	32
479 July 14	21	52	19	58	19	59	8	55	10	32	8	49	9	59
482 Mar 21	32	11	9	46	28	12	29	10	31	38	17	12	52	18
483 July 8	15	4	17	13	4	28	20	17	8	26	11	23	23	49
484 July 18	26	21	1	6	15	44	17	49	7	30	29	14	22	41
486 March 17	27	55	29	30	14	39	28	28	27	56	4	58	16	16
487 Sept 5	13	15	8	0	19	26	9	32	10	52	10	52	0	29
497 Oct 28	6	56	20	57	26	49	5	53	20	20	25	36	17	56

The modern-computed date for Table 2 in degrees of longitude, for the modern-equivalent year, Julian date and ephemeris time.

NICHOLAS KOLLERSTROM

Table 4 : Babylonian Charts

Year Date	Hour Babyl. local time	Mean S.Z.P. of planets (degrees)	S.D.	Ephem Time (hrs)
−234 June 3	4	7.6 ±	3.5	4,30
−199 June 4	4,50	6.2 ±	2.3	5,20
−198 Oct 31	6	9.1 ±	2.7	6,30
−87 Jan 5/6	0	4.3 ±	0.9	0 h.
−68 April 15	14,30	5.1 ±	2.7	14,30

GREEK CHARTS				
Year Date	Hour Alexan local time	Mean S.Z.P. of planets (degrees)	S.D.	Page in G.H
40 April 5	12	5.2 ±	1.8	80
46 Jan 3	22	1.3 ±	4.7	19
75 July 19	22	3.2 ±	4.1	89
76 Jan 24	6	3.5 ±	2.7	91
81 March 31	21	4.6 ±	2.4	24
110 March 15	19	3.4 ±	3	105
137 Dec 4	8	0.9 ±	2.8	41
260 Sept 29	15	0.2 ±	2.8	61
338 Dec 24	20	−0.2 ±	0.5	67
412 Feb 8	9	−0.3 ±	1.4	135
440 Sept 29	16	−0.7 ±	2.6	141
463 Apr 25	13	−1.8 ±	1.0	142
474 Oct 1	11	−2.4 ±	1.4	142
478 Aug 29	23	−2.9 ±	1.1	144
479 July 14	9	−2.9 ±	1.0	146
482 Mar 21	17	−2.5 ±	0.8	146
483 July 8	7	−0.6 ±	4.5	146
484 July 18	6	−2.9 ±	5.5	147
486 March 17	7.5	−2.9 ±	1.2	148
487 Sept 5	7	−4.7 ±	4.0	149
497 Oct 28	9	−2.2 ±	1.1	152

Mean longitude difference values sidereal-tropical per chart, obtained by subtracting the Table 3 values from the equivalent longitudes of Table 2 and adding 30° where necessary, then taking the mean and standard deviation of these difference-values per chart. The hour of local time is also given, with Ephemeris Time for Babylonian charts.

Difference in longitude
in charts from AD 40 to AD 497

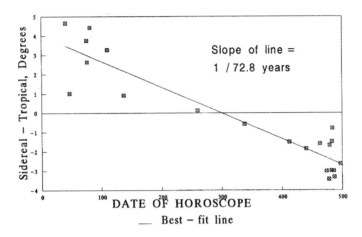

Figure 2. Graph of mean differences in celestial longitude between those
given in *Greek Horoscopes* and those computed using modern tropical longitudes, for 22 horoscopes. Each point is the mean of four to six values, viz.
Longitudes of the planets excluding Mercury, plus Sun and Moon. A best-fit
line has been put through the data.

Difference in longitude
in charts from AD 40 to AD 497

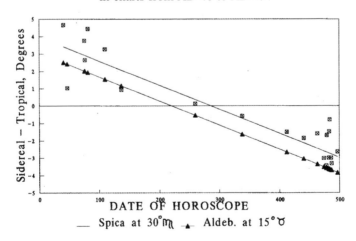

Figure 3. As before, with two lines put through these points of gradient one
degree per 72 years (rate of precession) to show the " expected " difference
values. They take Spica (α-Virgo) at 30° Virgo or Aldebaran (α-Taurus) at 15°
Taurus.

If the SZP values of this chart collection are indeed scattered about a single line, it would mean that astrologers were effectively using a single reference in this period, as was *sidereal*. Figure 3 shows two likely choices for such a sidereal zodiac position, defined by (a) positioning the star Aldebaran (α-Taurus) at 15° Taurus, or (b) Spica at 30° of the Virgin, ie close to the boundary. As we saw earlier, these references are roughly one degree apart. The graph indicates that the second of these is close and parallel to the best-fit line. This is similar (within 0°18') to having ß-Gemini define the 30° Gemini boundary, as was shown in Table I.

The first seven charts selected from Neugebauer's *Greek Horoscopes* used in Figure 2 span the period AD 40-140, and have an average SZP value of 3.7°±1.3°. For the group of the last 13 charts, spanning the period AD 410-500, the corresponding figures are - 1.9°±0.4°. These standard deviations give an estimate of increasing planetary ephemerides accuracy in antiquity. This appears as having increased several-fold between the first and fifth centuries. This comparison omits the two charts shown from the third and fourth centuries.

Planetary " error-values " were estimated within the two groups, as deviations from a sidereal trendline through the data. Sun and Moon positions had standard deviations of some two degrees while for the planets it was nearer three degrees, excepting Mercury which was over four degrees. These results accord with planetary error-patterns published by Owen Gingerich from Ptolemy's *Almagest*, showing that Ptolemy's Mercury-errors regularly went up to six or seven degrees[30].

How does the zodiac framework used by Babylonian horoscopes, in cuneiform, from the third to first centuries BC, compare with the later Hellenistic horoscopes on papyrus surveyed above ? Francesca Rochberg has collated in total twenty-eight such Chaldean horoscope tablets, spanning 410-69 BC[31], eight of which used degree longitudes to some extent. Five had sufficient to apply the above procedure, ie between four and six longitude values excluding that of Mercury. These five charts spanned 235 BC to 69 BC. Each of these charts had some time-of-day information, but less specific than for the Greek horoscopes, eg, one may be told which " part of the night " it was in, which planets were above the horizon and which had just set. Thus lunar longitudes may have an error of one degree or so. As regards the problem of converting from the Babylonian lunar-month calendar to the equivalent modern date, it

30. Ptolemy's Mercury theory generated regular errors of up to 6°, whereas for other planets his errors were only a degree or so: Owen Gingerich, *The Eye of Heaven*, New York, 1993, 67-68.
    31. F. Rochberg, " Babylonian Horoscopes ", *Trans. Amer. Phil. Soc.*, 1998, 88 part 1 ; F. Rochberg, " Babylonian Horoscopy : The Texts and their Relations ", *Dibner Institute Procceedings*, N. Swerdlow, Ed., 1998, M.I.T. Press, forthcoming. The 88 BC chart was previously published by Rochberg in " Babylonian Seasonal Hours ", *Centaurus*, 32 (1989), 146-170, 162 ; the 235 BC chart was published by E. Sachs in " Babylonian Horoscopes ", *Journal of Cuneiform Studies*, 6 (1952), 49-75, 60.

may suffice to quote Morrison and Stephenson, from their eclipse studies, that : " Experience has shown that high reliance can be placed upon the Babylonian dates "[32].

Figure 4 shows the same graph as before, with the Babylonian chart SZP values added. Only one " theoretical " line for zodiac position is used, that for the star Spica being placed at 30° of the Virgin, and the mean SZP value computed by Huber for 100 BC has been inserted as a point, the " Huber point ". A best-fit trendline has also been added, hardly distinguishable from the Spica line, whose slope was one degree per 74.0 years. This graph suggests that a single reference-framework for the sidereal zodiac was used by both Babylonian and Greek astrologers, enduring over some eight centuries, before being forgotten in the Dark Ages. The data here presented does support the above-quoted claim of Walker and Britton concerning β-Gemini, both for Greek charts (one might here prefer to use a term like, " Hellenistic ") as well as Babylonian.

Difference in longitude
in charts from BC 248 to AD 497

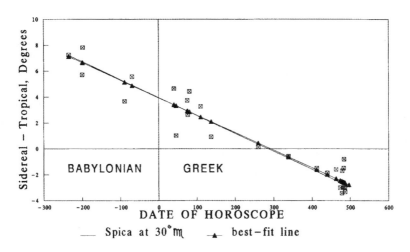

Figure 4. As before, but with only the Spica-reference line. The five early charts (from centuries BC) are Babylonian, while later ones are Hellenistic. A best-fit line has also been included.

Van der Waerden found that early Egyptian planetary tables were sidereal, in terms of their dates of planetary ingresses into zodiac signs. Three such

32. L.V. Morrison, F.R. Stephenson, " Contemporary... ", *op. cit.*, p.16 ; also F. Rochberg-Halton, " Babylonian Horoscopes and their Sources ", *Orientalia*, 58 (1989), 102-123.

tables had been partially preserved, of which two had a decent number of dates. For the reign of Augustus, Neugebauer had found that they were displaced on average + 4° with respect to modern values, which systematic difference decreased with time, and, reduced to the date of -100, the systematic difference became equal to 4 or 5 degrees. Concluded van der Waerden, " This means that the Egyptian mathematicians worked on the basis of a sidereal division of the ecliptic which almost coincided with the Babylonian division "[33].

Professor Pingree published a paper with the pessimistic title, " Hellenophilia versus the History of Science "[34]. The tendency to ignore the sidereal zodiac of antiquity, as having been the basis for measuring celestial longitude over nigh-on a thousand years, so that historians jump straight from the irregular-sized constellations of early Babylon to the tropical zodiac of the Greeks, is an example of this bias. For example, Owen Gingerich's essay " Origin of the Zodiac ", recently republished, gives its readers no clue that such a twelvefold sidereal framework ever existed[35], nor likewise do articles by Gurshtein in Sky and Telescope and Vistas in Astronomy[36] — thereby ignoring the basis of celestial longitude measurement in antiquity. John North's recent history mentions the topic, but only as a source of confusion whereby constellations and signs were liable to be muddled up[37].

The thesis here advocated can readily be tested. A further collection of horoscopes from antiquity is due to be published shortly[38]. Planetary tables of antiquity[39] and Babylonian Diaries that give planetary ingress dates[40] should both show more or less the same framework for longitude measurement (with the exception, possibly the sole exception, of Ptolemy's Handy Tables which were tropical). The thesis finds support from the fairly recent findings of Alexander Jones, as to how astronomical theory in Roman Egypt in the early centuries AD " in large part evolved from the predictive methods known to us from Babylonian tablets of the last four centuries BC "[41], and even that : " ...it

33. B.L. van der Waerden, Science Awakening II, Oxford 1974, 309.

34. D. Pingree, " Hellenophilia versus the History of Science, Isis, 83 (1992), 554-563.

35. O. Gingerich, " The Origin of the Zodiac ", Sky & Telescope, 67 (1984), 218-220, reprinted in The Great Copernicus Chase Gingerich, 1992, Ch. 2.

36. A. Gurshtein, " On the Origin of the Zodiacal Constellations ", Vistas in Astronomy, 36 (1993), 171-190.

37. J. North, Fontana, op. cit., 39.

38. A. Jones, Astronomical Papyri from Oxyrhynchus, Memoirs of American Philosophical Society, forthcoming, c.1999. Other papyrus horoscopes are collected in Donata Baccani, Oroscopi Greci, Messina 1992.

39. A. Jones, " Studies in the Astronomy of the Roman Period III Planetary Epoch Tables ", Centaurus, 40 (1998),.1-41.

40. See, eg., A.J. Sachs, " The Latest Dateable Cuneiform Tablets ", Kramer Anniversary Volume, Ed. B.Eichler et. al., 1976, Verlag, 379-398 ; A.J. Sachs, C.Walker, " Kepler's View of the Star of Bethlehem and the Babylonian Almanac for 7/6 B.C. ", Iraq, xlvi (1984), 43-55.

41. A. Jones, " Studies in the Astronomy of the Roman Period, I ", Centaurus 39 (1997), 1-36, 2.

is now clear that practically the whole of Babylonian planetary theory was current knowledge in Roman Egypt, well after the publication of Ptolemy's writings and tables "[42]. A primary motive for the studying of this matter concerns cultural continuity. If the same framework for measuring celestial longitude was being used north of the Mediterranean in centuries BC as south of the Mediterranean in the early centuries AD then this has vital implications for the transmission of astronomical science.

ACKNOWLEDGEMENTS

I am grateful to Robert Powell for permission to use the sidereal zodiac diagram of Figure 1, to John Britton and Christopher Walker for helpful advice, and to Anne Tihon for help on Table 2.

42. A. Jones, " Babylonian Astronomy and its Legacy ", *Bulletin of the Canadian Society for Mesopotamian Studies*, 32 (1997), 11-16, 16.

Astronomical Aspects of Robert Hooke's Scientific Research

Hideto NAKAJIMA

INTRODUCTION

Who was Robert Hooke ? This is an old and new question. Since the first academic paper appeared in 1913, historians of science have continued to ask the question. In 1989, in his paper published in *Robert Hooke : New Studies,* Steven Shapin had to raise the same question again[1].

A part of the reason why we fell into this situation seems to lie in the negligence of historians of science who did not pay enough attention to Hooke *per se*. For example, when I started my research on Hooke, even the name of his mother remained unknown. As a lucky Japanese, I discovered his father's last will accidentally, and could establish Hooke's family tree[2].

Another reason why we ask the same question repeatedly may be the lack of paradigm to handle his scientific career consistently. The aim of this paper is to propose a framework with which we will be able to reach better understanding on Hooke. I shall emphasize here the importance to see Hooke as an " astronomer ".

PRECEDING IMAGE OF ROBERT HOOKE

Who was Robert Hooke ? The most common image of him must be that of *Micrographia* (1665) and of *Spring* (1678), two of his major works. Hooke's Law (the extension of springs is proportional to the force exerted) appears on high school textbooks for physics. Hooke is usually mentioned in textbooks for

---

1. S. Shapin, " Who was Robert Hooke ? ", in M. Hunter and S. Schaffer (eds), *Robert Hooke : New Studies*, Woodbridge, Boydell, 1989, 253-285.
2. H. Nakajima, " Robert Hooke's Family and his Youth ", *Notes Rec. Roy. Soc. Lond.,* 41 (1984), 261-278.

biology also, because he is said to have discovered the cell structure of the plants.

Historians of science, of course, know much more about Hooke. He was a member of the Oxford group. He was an able assistant of Robert Boyle, and Hooke produced the air pump with which Boyle discovered Boyle's Law. After the Restoration, Hooke was elected to the first curator of the Royal Society of London. His demonstrating experiments enchanted the fellows of the Society. Hooke's institutional activities and his private life were uncovered by preceding research from Margaret 'Espinasse (1956)[3] to S. Shapin (1989).

Who was Robert Hooke as a " scientist " then ? In comparison with the analysis of his social life, the research on Hooke as a scientist does not seem to be improved well. We are still haunted by the traditional image that Hooke was a cause of headache of Newton.

As the late Westfall delineated, when Newton made his debut in 1672 with his reflecting telescope, Hooke vehemently criticized Newton's reflector and Newton's new theory about light and colors[4]. In the field of mechanics, Hooke is known to have insisted his priority for the theory of universal gravitation, when Newton was preparing the draft for his *Principia*. In short, Hooke was recognized as the enemy of Sir Isaac Newton.

ICONOGRAPHICAL ANALYSIS OF ROBERT HOOKE

Who was Robert Hooke as a scientist I would like to give a new answer to the question by an iconographical analysis.

In their work[5], Steven Shapin and Simon Schaffer cited the frontispiece of Sprat's *History of the Royal Society* (1667), which was published to protect the Royal Society against criticisms from the public.

Shapin and Schaffer emphasized the importance of Boyle's air pump in the early stage of the Royal Society. They believed that its importance was the reason why his pump appeared in the frontispiece (figure 1). They wrote " the powerfully emblematic status of the air-pump is manifested in its contemporary iconography "… " No new device had taken the place of the *machina Boyleana* as an emblem of the Royal Society's experimental programme "[6].

If this is true, this drawings verify the importance of not only Boyle's pump but also Hooke's astronomical research in the early Royal Society. When you look at the enlarged version of the frontispiece, you will see a tall telescope with bizarre outfit gentleman behind Boyle's air pump. We can find out the

3. M. 'Espinasse, *Robert Hooke,* London, Heinemann, 1956.

4. For example, see R.S. Westfall, *Never at Rest : A Biography of Isaac Newton*, Cambridge, Cambridge UP, 1980, Chap. 7 and Chap. 10.

5. S. Shapin, S. Schaffer, *Leviathan and the Air-Pump*, Princeton, Princeton U.P., 1985.

6. *Ibidem*, 32.

same telescope in Hooke's letter in 1667 to Hevelius, famous astronomer in Danzig (figure 2). Hooke wrote this picture to show Hevelius the way how he suspended his long telescope. I would like to argue that his telescope was drawn on Sprat's book because of its importance.

Hooke's astronomical activities can be traced in his contributions to the *Philosophical Transactions*. Hooke published thirty-seven papers in the *Philosophical Transactions*. Eighteen of them, nearly 50 %, are related to telescopes and telescopic observations.

Iconographical analysis and an analysis of these papers in the *Philosophical Transactions* clearly suggest that Hooke was active in the field of astronomy.

HOOKE'S ASTRONOMICAL RESEARCH

As van Helden wrote in his papers, astronomers after Galileo were interested in refining his discoveries[7]. They are interested in the observations of the surface of the moon, of the strange appearance of Saturn, and of minutes structures and satellites of the planets. For example, Hevelius wrote a book on Selenography. Christiaan Huygens identified the ring of Saturn. Cassini was interested in diurnal rotation of Mars.

These researchers applied long telescopes for their observations. Though Descartes and Kepler demonstrated theoretically that telescopes with conical lenses were required to improve telescopes, actual development followed a different way. Astronomers started to extend the telescope longer and longer, because this was an alternative way to avoid spherical aberration.

Hooke's astronomy kept the line with the contemporary research trend. Hooke usually used long telescopes for astronomical observations as others did. As far as I found in his works and letters, he referred to 36 feet (about 10 meters) and 60 feet (about 20 meters) telescopes.

With these telescopes, he made a small contribution to selenography in his *Micrographia*[8]. In 1666, results of his observations were printed in the first volume of the *Philosophical Transactions*. He wrote a detailed picture of Jupiter and Saturn. By the periodic change of the figure of Mars, Hooke concluded that the period of diurnal rotation should be about twenty-four hours.

Hooke's observations of planets gave strong impression to Hevelius. In 1667, Hevelius ordered Hooke to produce a 60 feet telescope for him. Hooke shipped the telescope two years later. Hevelius did not hesitate to praise the high quality of the telescope. In 1673, Hevelius published a 60 feet telescope in his *Machina coelestis, pars prior*. This seems the one Hooke sent to him,

---

7. For example, see A. van Helden, " The Telescope in the Seventeenth Century ", *Isis,* 65 (1974), 38-58.

8. See, pp. 241 ff.

because Hevelius wrote that he set this telescope according to English method[9].

Though both Hooke and Hevelius preferred to use long telescopes, their attitude to telescopic sight and eyepiece micrometer was utterly opposite. Hooke believed that these instruments were essential for precise observations. Hevelius, however, stressed that naked eyes fitted better for the purpose. They started to dispute about the problem, and their controversy arrived its peak when Hooke published one of his Cutler Lectures titled *Some Animadversions on the First Part of Hevelius His Machina Coelestis* (1674).

Hooke himself applied a telescopic sight and an eyepiece micrometer to an astronomical observation. He set them into a zenith scope and tried to find out parallax of a fixed star. He published the result of this observation in another Cutler Lecture titled *An Attempt to Prove the Motion of the Earth* (1674).

Historical meaning of his work of *Helioscopes* (1676), a third Cutler Lecture, could be easily grasped if we see Hooke from the direction of astronomy. In February 1667, he showed the Royal Society a contrivance " to contract the power of a long telescope into a short one ". Long telescopes were difficult to handle. To solve the problem, he struck to an idea to fold telescopes utilizing mirrors. But, because of low reflective power of mirrors, Hooke could not get bright images of stars. So he converted the instrument to helioscopes for which dark image was rather preferable.

CONCLUSION

As I discussed, many of Hooke's researches are understood consistently, if we see him as an astronomer. Of course, his activities were not restricted to astronomy. But I would like to argue that my methodology, to see Hooke as an astronomer, will broaden our perspectives, and will open new research programs to embed him into his historical context.

---

9. Hevelius (1673), p. 392.

FIGURE 1

FIGURE 2

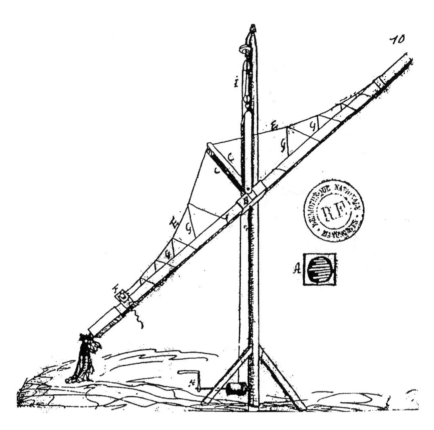

Hooke's long telescope, 1667

DIALOGUE CONCERNING THE TWO ASTROLOGIES IN COIMBRA JESUIT'S THOUGHT[1]

Luís Miguel CAROLINO

The dialogue concerning the scope and limits of astrology, that is the influence of the celestial region on the sublunar region was greatly revived during the Renaissance. The contact with the Greek originals, particularly with Plato, with the Neo-Pythagoreans and especially with the Neo-Platonists, gave rise to a movement of criticism and detraction of the judicial astrology, apparent in authors such as Ficino and Pico della Mirandola among others[2].

Thus, from the 14th to the 16th century, astrology came into a " strange limbo " phase, in the words of John North. If, on the one hand, the theological stances about the relation celestial region/sublunar region were still accepted without hesitation, on the other one could see an attempt to rationalise the astrological theses[3].

This attempt to rationalise finds its way among the scholastics of the Renaissance. Therefore, along with theoretical statements and fundaments of the relation between the celestial region and the terrestrial where it traditionally stopped[4], Scholasticism starts to include questions that were strange to it until then, namely the discussion about the validity of astrological prognostication. From the examination of a detailed " Catalogue of Questions on Medieval Cosmology, 1200-1687 ", made by Edward Grant[5], we may see that

1. Work carried out with the support of Biblioteca Nacional de Lisboa.

2. See E. Garin, *O Zodíaco da Vida. A polémica sobre a astrologia do século XIV ao século XVI*, Lisboa, Editorial Estampa, 1988, 75-101 ; S.J. Tester, *A history of western astrology,* Woodbridge/Suffolk, Boydell Press, 1987, 204-243.

3. *Cf.* J. North, *The Fontana history of astronomy and cosmology*, London, Fontana Press, 1994, 260.

4. See E. Grant, " Medieval and Renaissance scholastic conceptions of the influence of the celestial region on the terrestrial ", *Journal of Medieval and Renaissance Studies*, 17 (1987), 1-23 ; E. Grant, *Planets, stars and orbs. The medieval cosmos, 1200-1687*, Cambridge, Cambridge University Press, 1996, 569-617.

5. E. Grant, *Planets, stars and orbs...*, 681-741.

questions like " What do celestial bodies especially effect with regard to humans, and what can be predicted from them [that is, celestial bodies] concerning human affairs ? " or " Whether by means of the influences of the stars an astronomer can predict the future by the influence which the heaven exerts on sensitive and intellective powers " mainly developed after the 16th century and during the next.

If it is true that nearly all the stances deny the validity of judicial astrology, it is also significant that on the stage of scholasticism there appear players who were traditionally outside its scope and acted in other cultural plans : the astrologers. In the exercise of denying judicial astrology, developed in questions like " Whether astrologers can predict effects that are dependent on observation of the stars ", scholastics have started to focus on the figure of the astrologer, describing the range of practices and services provided by them.

Renaissance commentaries to Aristotle's *De Coelo* formulated by the Jesuits of the Colégio das Artes de Coimbra (Portugal) in the transition from the 16th to the 17th century — *Commentarii Collegii Conimbricensis Societatis Iesu in quatuor libros De Coelo Aristotelis Stagiritae*[6] — give evidence of this new relation with astrology. These commentaries were the base of the pre-college teaching provided for at Colégio das Artes, inaugurated by John III in 1542. The *De Coelo* under investigation here, printed in 1593, is a scholarship work[7], studied in the third of the four years' course before entering the University of Coimbra, during the 16th and 17th centuries.

It is this new approach to astrology that we shall analyse on the next pages. Firstly, we shall explore the perspective of the Conimbricenses on the influence of the celestial region on the sublunar region ; then, we will explore the Conimbricenses response to astrological prognostication, stressing, in addition to the arguments about its illicit nature, the contours drawn by astrologers and their practices.

INFLUENCE OF CELESTIAL BODIES ON THE TERRESTRIAL REGION

As it was unanimously accepted by scholasticism[8], also the Conimbricenses deemed incontrovertible the influence of celestial bodies on the sublunar region. They believed this relation to be so evident that it was not even necessary to resort to the arguments of the scholars to prove it, because the evident

6. See Gomes, 1992 ; Lohr, 1988, 98-99 ; Martins, 1989 ; Jesué Pinharanda Gomes, *Os Conimbricenses*, Lisboa, Instituto de Cultura e Língua Portuguesa, 1992 ; C.H. Lohr, *Latin Aristotle Commentaries, II : Renaissance Authors*, Florence, Olschki, 1988 ; A. Manuel Martins, " Conimbricenses " in *Logos - Enciclopédia Luso-Brasileira de Filosofia*, Lisboa, São Paulo, Editorial Verbo, 1989, vol. 1, 1112-1126.

7. Although it is a known fact that its author was a Jesuit named Manuel de Góis (1542-1597), this work is by its intentionality, materialised for instance in the express absence of the author's name, a work of a collegial nature. See Gomes, *op. cit.*, 51-63.

8. See E. Grant, *op. cit.*, 570.

experience taught it[9].

Thus, to the examples of celestial causation accumulated throughout centuries of scholastic thought[10], the Coimbra Jesuits added some examples that made unquestionable the theory of the astrological relation. Thus, they ascribed to the Sun the division of the year into four seasons. According to them, the motion of the Sun produced a quadripartite division of the year which encompassed the two equinoxes of spring and autumn and the solstices of summer and winter. This induced the alternation of heat and cold, originating the generation and destruction of things[11].

With a powerful belief in the causal relation between the celestial world and the terrestrial, the discussion about the agents that operated that relation was long held by the Conimbricenses and scholastics in general[12]. The instrumentalities of celestial action were generally identified as motion, light, and influence (*influentia* or *influxus*). This influence materialised in the sublunar region by the production of the four primary qualities — hotness, coldness, wetness, and dryness — which, combined among themselves, particularly in the case of hotness and coldness, produced generation and corruption in the sublunar sphere, that is terrestrial life.

But as to the operation of the celestial causes, if the first two categories — motion and light — gathered a large consensus, the category of influence was the subject of a heated debate in the universe of scholastic philosophy[13]. Therefore, the third question[14], as it may be immediately inferred by its title — *Utrum corpora coelestia interuentu motus et lucis duntaxat ; an etiam per alias occultas qualitates influant* — assumed the existence of this third celestial agent as its main purpose.

9. *Sed neque sapientum testimoniis opus est ad id comprobandum, quod manifesta docet experientia, Commentarii Collegii Conimbricensis Societatis Iesu in quatuor libros De Coelo Aristotelis Stagiritae,* Olisipone, Ex officina Simonis Lopesii, 1593, lib. 2, cap. 3, qu. 1, art. 2, p. 156.

10. Examples that explained, as it may be seen in the Conimbricenses *De Coelo,* the influence of the Sun through phenomena like the heliotropism, (...) *sol Heliotropii, et Scorpiuri flores ab ortu ad occasum secum vertit, De Coelo,* lib. 2, cap. 3, qu. 1, art. 2, p. 157, and the crowing of the cock daily before sunrise, *idemque [Sol] peculiarem vim gallo infundit, ut quidam autumant ; atque eam causam esse credunt, cur gallus ad mediam noctem canit, De Coelo,* lib. 2, cap. 3, qu. 1, art. 2, p. 157 ; and the Moon through the association of its motion with the tides ; the opening and closing of oysters and other shells ; and the behaviour of certain animals such as the owl and the ant, *Eius quippe impulsu aestus maris ultro, citroque reciprocatur. Ostrea, conchilia, et conchae omnes cum ea pariter incrementa, pariter decrementa accipiunt : atque eius numero respondent soricum fibrae. Formica nono die ab eius coitu, quem sibi aduersum nouit, nunquam e latebris exit, interlunio semper quiescit, plenilunio etiam noctibus in opus incumbit, De Coelo,* lib. 2, cap. 3, qu. 1, art. 2, p. 157.

11. *(...) videmus solis motum afferre quadripartitam anni distinctionem, quae duo aequinoctia vere et autumno, ac totidem solstitia aestate et hyeme complectitur. Unde calorum et frigorum vicissitudo, et rerum generatio obitusque existit, De Coelo,* lib. 2, cap. 3, qu. 1, art. 2, p. 156. Example already cited in E. Grant, 1987, 7 ; E. Grant, *Planets, stars and orbs...,* 576.

12. See E. Grant, " Medieval and Renaissance scholastic conceptions... ", 9-15 ; E. Grant, *Planets, stars and orbs...,* 586-615.

13. On the different positions about the celestial causative agents influencing the terrestrial region, see Grant, *Planets, stars and orbs...,* 586-615.

14. *De Coelo,* lib. 2, cap. 3, qu. 3.

Let us start with the origin of hotness. In *De Coelo* two explanations were given, often proposed by scholastics, which focused on the motion of celestial bodies. On the one hand, assuming the non-existence of vacuum, it was advocated that the Moon, in its continual rotation, moved the element fire (*ignis*), which in turn heated the air that was contiguous[15] and pervaded the space as far as the earth's surface. On the other, it was thought that the motion of the celestial bodies, particularly the Sun which generated the seasons as we have seen, allowed a variety of changes and generations when it was closer to or farther from the earth[16]. The motion of celestial bodies appears thus as the principal means through which the celestial region affects the sublunar region. However, on a par with this celestial agent, it was said in *De Coelo* that light could also produce the same effect, that is, not only did it illuminate the inferior bodies but it also heated them[17].

More complex, as it appears in the Conimbricenses text, was the origin of coldness. In essence, its authors tended to deny that light might generate coldness. Reflecting the long and intense controversy that agitated scholastic circles[18], they considered that celestial agents such as light could not cause hotness and coldness at the same time, because these were contrary qualities. Thus, because heat was the result of magnanimous light, it was therefore improbable that the origin of two contrasting phenomena or opposing qualities such as hotness and coldness, or wetness and dryness, could be found in this celestial agent. Therefore, light could not produce coldness[19].

15. *globus lunae perpetua circuitione ignem rapit ; ignis vero cohaerentem sibi aerem, qui agitatus mutuo partium attritu incalescit, De Coelo*, lib. 2, cap. 3, qu. 3, art. 2, p. 163.

16. *motus admouet, remouetque a nobis planetas, quorum, praesertim solis, acessu, et recessu fit quadripartita anni varietas cum illa, quam supra diximus, alterationum, generationumque varietate, De Coelo*, lib. 2, cap. 3, qu. 3, art. 2, p. 163.

17. *cum nemo sit, qui non videat experiaturque lucem e coelo fusam non solum collustrare, sed calefacere etiam haec inferiora corpora, De Coelo*, lib. 2, cap. 3, qu. 3, art. 2, p. 163

18. See E. Grant, *op. cit.*, 603-605.

19. *Quo fit vt quemadmodum elementaris calor frigori aduersatur ; ita et coelestis, et ex consequenti vt coelestis nequeat frigiditatem gignere, alioqui vnum contrarium aliud produceret : cum tamen contrariorum ea conditio et natura sit, vt se se mutuo interimant. Item quia licet coelestis calor sit generosa lucis soboles, non proinde verisimile est, duplicem illam tam dissimilium, ac repugnantium qualitatum coniugationem, caloris et frigoris, humoris et siccitatis, virtute in eo contineri, De Coelo*, liv. 2, cap. 3, qu. 3, art. 2, p. 164. The Conimbricenses were also opposed to the theory that assumed the existence of a unique light that could cause not only hotness but also coldness and the other qualities, *Alij igitur contendunt solam lucem praedictas qualitates parere, non interuentu caloris a se geniti, vt aiebat Picus, sed immediate prout ipsamet huius, aut illius sideris instrumentum fit, vt tertium argumentum astruere nitebatur. Hoc etiam placitum hisce rationibus confutatur, De Coelo*, lib. 2, cap. 3, qu. 3, art. 2, p. 164, and they also rejected the possibility that some stars could emit light that was cold-producing. This theory, often posed in scholastic manuals, was widely considered as unacceptable E. Grant, *op. cit*, 604-605. Coimbra Jesuits also rejected that the concept of coldness advocated by Averroës among others — which explained coldness by the absence of heat or its insufficiency — could ground the thesis that hotness was produced directly by the celestial agents, while the other qualities — coldness, dryness, and wetness — were produced by accident or indirectly, *Namque, vt Thomas recte disputat, si corpora coelestia ea tantum ratione frigiditatem gignerent, quia causam caloris effectricem remouent ; vel vt etiam eorum nonnulli interpretantur, quia remisse calefaciunt, sequitur alias qualitates, excepto calore, non per se, sed ex accidente, a coelo effici ; quod falsum est, De Coelo*, lib. 2, cap. 3, qu. 3, art. 2, p. 165.

Thus, the explanation for the origin of coldness would not be in the operation of these two celestial agents but in the existence of another celestial agent potentially more efficient in the production of the other qualities — coldness, dryness, and wetness — but also more difficult to be established : the influence (*influentia* or *influxus*)[20].

After rejecting the theory that motion and light could produce coldness, the Conimbricenses described this occult force (*vis occulta*) or influence. Through this force, acting in an imperceptible way, a wide range of effects could be explained, which from the Aristotelian standpoint could hardly be explained otherwise, particularly coldness, we presume. Thus, in addition to the examples generally cited by scholastics and also repeated in *De Coelo* — the magnetism[21] and the ebb and flow of the tides[22] — the Conimbricenses provided other " experiences ", or examples. In their opinion, celestial bodies generated through occult forces or influences, gold and other metals in the bowels of the earth, where light or motion could not penetrate[23].

The existence of this agent of celestial causation — influence (*influentia* or *influxus*) — conforms, as we have seen, to the theory where the third question is based, which is the main conclusion of the account — the *conclusio disputationis* — that is the theory that celestial bodies, which are illuminated,

20. *Coelum non interuentu solius motus, et lucis, sed alias etiam occultas vires, quas influentias vocant, in sublunarem mundum influit, De Coelo*, lib. 2, cap. 3, qu. 3, art. 2, p. 163. Other celestial agent referred to is universal cause (*causa universalis*). This, which is deeply grounded by scholastic philosophy from the 13th to the 17th centuries, explained the preservation of the world in its own order — namely by preventing the possibility of vacuum formation — and thus it was considered by some philosophers like Suarez as God's way of preventing the world from destruction (Grant, *op. cit*, 616). In *De Coelo*, not only this role as preserver of the cosmos unity was referred to, but rather primary emphasis was placed on the ability of the universal cause to produce all the primary qualities, thus objectifying the intervention of the celestial region in the terrestrial reality, *oportuit causam vniversalem, cuiusmodi est coelum, per se atque ex instituto producere omnes primarias qualitates, vtpote operum naturae ministras, ac coelestis formae vicarias, De Coelo*, lib. 2, cap. 3, qu. 3, art. 2, p. 165.

21. *Rursum acus attritu magnetis confricata trahitur a coeli partibus polo vicinis vt in 7 Physicae auscultationis libro disseruimus, non vehi autem lumine eiusmodi attractoriam vim argumento est, quod etiam densa caligine, et in quolibet tenebricoso loco, acus versus illum coeli tractum obuertitur. Negari igitur non debet id alterius virtute ope fieri, De Coelo*, lib. 2, cap. 3, qu. 3, art. 2, p. 164.

22. *Item fluxus et refluxus Oceani fit a luna, vt communi philosophorum consensione in Meteoris ostendemus : fit autem saepe non apparente luna in nostro hemisphaerio : nec vllam ad ipsum lucem effundente. Igitur luna aliam obtinet vim, per quam id efficiat. Praeterea experientia compertum est lunam, dum cum sole coit, vehementius mouere haec inferiora : et in morbidis corporibus saepe numero acriorem doloris vim efficere, maioresque assultus Oceani, et vndarum reciprocationes excitare : et tamen in eo congressu multo minorem lucem, possidet : habet igitur occultam aliam facultatem a luce distinctam, per quam operatur, De Coelo*, lib. 2, cap. 3, qu. 3, art. 2, p. 164.

23. *Multa a coelo perficiuntur non intercedente motu, aut luce : igitur coelum non sola luce motuve : sed alia etiam occulta vi ad agendum pollet. Probatur assumptum, quia astra procreant aurum, caeteraque metalla in terrae gremio , quod neque motus attingit, neque lux permeat : neque enim bruta illa terrae crassities radio penetratur, De Coelo*, lib. 2, cap. 3, qu. 3, art. 2, p. 164.

produce heat in the inferior world, although other forces exist that produce different qualities ; those forces are called influences[24].

JUDICIAL ASTROLOGY

Stated and established the causal relation between the celestial region and the sublunar reality, an important question was posed to the Renaissance scholasticism — to find out whether the observation and study of celestial bodies enabled the prognostication of future earthly events or possible human actions, and also the possibility to learn about the personality of each human being. Thus, the practices of judicial astrology, which have been secularly strange to scholastics, became a subject of debate.

Differentiating between two hypothetical ways by which celestial bodies could act on terrestrial bodies, directly — i.e. inducing something against men's will, or indirectly — i.e. when celestial bodies, remotely and by accident, could act on the will through the organs of the human body and their faculties[25], the Conimbricenses, in line with the current of Renaissance Aristotelians, were categorical on this issue : celestial bodies could act indirectly on the will, but not directly[26].

Therefore, they left human will out of the causal relation celestial region/ sublunar region, which had been the key to the opposition to astrology since the first Fathers of the Church, because free will was the primary element of human responsibility before the Church and God[27]. This is also suggested by the theory that if human actions were governed by the influence of the heaven, they would not be free but natural, men would not act as human beings but as animals. Consequently, there would be no reason to choose, no guilt, no merit[28].

24. *Maneat igitur corpora coelestia, qua lucida sunt, in inferiori mundo calorem gignere, sed habere praeterea vires alias aliarum qualitatum effectrices, quae influentiae vocantur, De Coelo,* lib. 2, cap. 3, qu. 3, art. 2, p. 165.

25. *Ad explicationem huiusce dubij praenotandum est, bifariam intelligi posse corpora coelestia mouere voluntatem, nimirum directa, vel indirecta motione. Directa, idest, imprimendo per se, atque ; immediate aliquid in ipsam voluntatem ; indirecta, id est, remote et ex accidente eam inclinando interuentu organorum corporis et potentiarum eis inhaerentium (...), De Coelo,* lib. 2, cap. 3, qu. 8, art. 2, p. 184.

26. *Corpora coelestia possunt agere in voluntatem indirecta motione, directa non possunt, De Coelo,* lib. 2, cap. 3, qu. 8, art. 2, p. 184.

27. See especially : M.L.W. Laistner, " The western church and astrology during the Early Middle Ages ", *The Harvard Theological Review,* 34 (1941), 251-275. See also : E. Grant, *op. cit,* 569-570 ; S. Hutin, *História da astrologia,* Lisboa, Edições 70, s.d., 82 and 106 ; L. Ackerman Smoller, *History, prophecy and the stars. The christian astrology of Pierre d' Ailly. 1350-1420,* Princeton, New Jersey, Princeton University Press, 1994, 25-29.

28. *(...) si humanae actiones coelesti vi administrarentur, non essent liberae, sed naturales ; nec homo ageret, sed belluarum more ageretur, nullaque superesset ratio consilij, nulla culpa, aut meritum, De Coelo,* lib. 2, cap. 3, qu. 8, art. 2, p. 184.

Judicial astrology was thus imbued with a spirit of illegitimacy. However, the new cultural momentum of the Renaissance required an extended effort to depreciate judicial astrology. It was with this spirit that in *De Coelo* the analysis was extended not only to the qualifications of the astrologers, but also to the different practices used in the prediction process.

As to the qualifications, right from the first article on this subject where this issue and the position of divinatory astrologers were discussed, it was concluded that, according to the Conimbricenses, the services provided by an astrologer would range from the prediction of births and the knowledge of human nature beforehand, to luck, events, actions, future contingencies and other things that rely on free will[29].

Astrological practices were subject to a more comprehensive analysis, providing a wide range of divinatory procedures. The most common and widespread would associate planets and celestial constellations with the zodiacal house of birth in order to determine a person's character[30]. Thus, conjoining these factors, astrologers would conclude, according to the Conimbricenses, that those men who were benignly " looked " at by Capricorn were born kings ; by Aquarius, fishermen ; by Mercury, bankers ; by Orion, hunters ; by Mars, murderers. It was also predicted that persons born with ascendant in Gemini, being Saturn and Mercury together under Aquarius in the 9th region of the heaven, would become poets. Moreover, it was said that those born with Saturn in the zodiacal house of Leo would be fortunate and protected from calamities[31]. Concurrently with this traditional practice, other divinatory methods equally secular were mentioned such as palmistry and physiognomy — birthmarks were believed to be printed by the heaven making it possible to learn about a person's inclinations, virtues and vices[32] — and the resort to premonitory dreams[33]. Astrologers were also believed to have a conception of the

29. *Quod ergo ij, quos Genethliacos vocant, qui videlicet natalitiam diuinationem profitentur, cuiusque hominis mores, fortunam, euentus, actiones, etiam quae a libero arbitrio dependent, ac caetera futura contingentia certo praenuntiare valeant videtur probari posse, De Coelo*, lib. 2, cap. 3, qu. 9, art. 1, p. 186.

30. *Nam vt a moribus ordiamur, facile est Genethliacis notare in quo quisque sydere, quoue caeli statu conceptus fuerit, in quo natus horoscopo, et inde eius temperamentum elicere, De Coelo*, lib. 2, cap. 3, qu. 9, art. 1, p. 186.

31. *Obseruatum ab astrologis est eos, quos Capricornus benigne aspexerit, nasci reges ; quos Aquarius, piscatores ; quos Mercurius, Trapezitas ; quos Orion, venatores ; quos Mars, homicidas, et gladio casuros. Item ascendente Geminorum sydere, et Saturno, Mercurioque sub Aquario iunctis in nona coeli plaga, nasci vates. Praeterea qui Saturnum in Leone feliciter constitutum in genesi sortiti fuerint, eos multi calamitatibus ereptum iri ; et alia eiusmodi, quae ipsi notata habent, De Coelo*, lib. 2, cap. 3, qu. 9, art. 1, p. 187.

32. *Physiognomia ex lineamentis manuum, ex vultu, et totius corporis figura, tanquam e signis quibusdam a coelo impressis cognoscit propensionem et virtutes ac vitia cuiusque, De Coelo*, lib. 2, cap. 3, qu. 9, art. 1, p. 186.

33. *Nonnunquam dormientes, qui rationis vsu, et iudicio carent, ex afflatu sydereo futura quaedam praeuident, ac diuinant, vt testatur experientia. Ergo vigilantissima, subtilissimaque Astrologorum observatio ex suae artis praeceptis multo maiori iure futura praenoscere, ac praenuntiare poterit, De Coelo*, lib. 2, cap. 3, qu. 9, art. 1, p. 187.

heaven as a divine book where future events were written through marks that astrologers could naturally read[34]. Finally, some believed that demons had a natural disposition to predict future events[35].

To the Conimbricenses these practices were not worthy of any credibility, and they generally considered them as belonging to the realm of fantasy[36]. Thus, the conclusions drawn by astrologers from the conjunction of planets, constellations and zodiac were considered by the Conimbricenses as inventions with which astrologers filled the heavens as much as poets did with fables[37].

After the thorough study carried out by the Conimbricenses, two incontrovertible conclusions emerge : the life of terrestrial beings is strongly influenced by celestial bodies, however the inferences drawn by the superstitious traditional astrology do not regard the scope of such influence.

This conclusion was determinative of the position of the Conimbricenses towards prophecies, so popular in the society of their time, particularly with respect to the differentiation between the prognostications deemed legitimate and the illegitimate. But after this meditation, the distinction looked simple and natural.

In the first place, one could legitimately predict with exactitude and in advance the aurora, the twilight, the aspect and the eclipses of stars and planets, because those physical phenomena were based on the regularity and immutability of the celestial bodies' motion, which is assured by God[38].

Also the natural phenomena occurred in the sublunar region by the influence of celestial bodies, such as illnesses, coldness, storms, sterility, rains, droughts

34. *saltem sunt eorum signa, estque coelum quasi diuinus quidam liber, in quo scripta sunt, quae progressu temporum euenient. Igitur qui huiusce libri characteres, significationemque tenuerit, quam se tenere Astrologi profitentur, quaelibet futura, licet in se dubia, et incerta existant, certo cognoscet, De Coelo*, lib. 2, cap. 3, qu. 9, art. 1, p. 187.

35. *Daemones praecognoscunt futura contigentia (...), De Coelo*, lib. 2, cap. 3, qu. 9, art. 1, p. 187.

36. From the criticism to the interpretation of celestial signs were naturally exempt the signs which the Bible refers, such as the announcement of the Judgement Day made by God to men — through appalling changes in the weather — and the birth of Christ — through a star that guided the three Magi. *Illa autem verba Esaiae cap. 34 aliam habent intelligentiam, nimirum tantam futuram oppressionem earum gentium, quibus eo loco Deus vindictam comminatur, vt eis caelestia lumina extingui, et coelum complicari, ac recedere videri possit. Locus vero Apocalypsis 6. significat die iudicij coelum adeo tenebris obducendum, vt quemadmodum literae in libro inuoluto non videntur, ita nec sol, aut reliqua astra videri queant. (...) Quod vero attinet ad locum Geneseos, dicendum stellas fuisse in coelo positas vt essent in signa (...), De Coelo*, lib. 2, cap. 3, qu. 9, art. 3, p. 195.

37. *Figmenta sunt haec Astrologorum, qui fabulas non minus, quam poetae coelum impleuere, De Coelo*, lib. 2, cap. 3, qu. 9, art. 3, p. 194.

38. *Syderum planetarumque omnium ortus, obitus, aspectus, et eclypses, aliaque id genus ad coeli statum pertinentia possunt Astrologi certo, multoque antea praenoscere. Probatur, quia haec omnia tantum pendent a coelestium sphaerarum motibus, quos rato, indeflexo, et aequabili ordine fieri constat : Astrologi autem possunt eiusmodi motuum leges comprehensas habere. Erit tamen ea certitudo physica duntaxat, quia in diuina potestate est talem ordinem (si libuerit) immutare ; id quod non semel iam accidisse ex sacris literis constat, De Coelo*, lib. 2, cap. 3, qu. 9, art. 2, p. 188.

and earthquakes could be predicted by astrologers with great probability but could not be announced with certainty[39]. This impossibility resulted from the infinite complexity and variety of effects produced by all celestial bodies and which no human being could know in their entirety[40]. However, following Avicenna's example, the observation of the sky should be maintained but it was equally important to explore the nature of the inferior world, the different " stages " of the lands, the natural temperature of each region and all the near causes that with the concurrence of the heaven interfere with the course of terrestrial life[41].

Finally, following Saint Thomas Aquinas, all prognostication about human actions dependent on human will, casual successes or fortuitous events were naturally deemed illegitimate[42].

In brief, the text of the Conimbricenses indicates a clear attempt to characterise and combat judicial astrology. Not only is it denied, for theological reasons, the direct influence of celestial bodies on human behaviour, but critical remarks are also made to the astrological methodology. The methods used by astrologers were considered as largely arbitrary, casuistic, lacking accuracy, hence legitimacy. They gave way to accusations of superstition and charlatanism with which the astrological tradition was secularly struck by the Christian orthodoxy[43].

Thus, if it is true, as Edward Grant has said, that judicial astrology played very little role in scholastic natural philosophy[44], it should be stressed both the effort to depreciate judicial astrology and the detailed knowledge of the differ-

---

39. *Morbos, frigora, tempestates, sterilitatem, pluuias, siccitatem, terrae motus, aliaque eiusmodi naturalia effecta, quae infra lunam e coelestium corporum defluxu magna ex parte obueniunt, possunt Astrologi admodum probabiliter, non tamen certo enuntiare*, De Coelo, lib. 2, cap. 3, qu. 9, art. 2, p. 189.

40. *(...) quia per difficile est omnium syderum, a quibus eiusmodi euenta inhiberi queunt, vim et concursum ad amussim tenere : nec poterit non confundi ratio disciplinae in tanta stellarum multitudine, et influendi variatate, quae nulli mortalium nota est*, De Coelo, lib. 2, cap. 3, qu. 9, art. 2, p. 189.

41. *Praeterea, vt Auicenna in sua Metaphysica aduertit, ad eiusmodi effectorum cognitionem non sufficit Astrologo coelum inspicere, sed oportet inferioris mundi naturam, diuersos terrarum situs, et innatam cuique regioni temperiem, ac denique omnes proximas caussas, quae ad promouendum aut impediendum coeli concursum aliquid momenti habent, exploratas habere*, De Coelo, lib. 2, cap. 3, qu. 9, art. 2, p. 189.

42. *Neque actiones, quae ex humana voluntate pendent, neque successus contingentes, aut casus fortuitos possunt Astrologi praenoscere. Haec conclusio est sanctorum patrum, et Theologorum communis, probaturque a D. Thomas 2.2.q.95. art. 5*, De Coelo, liv. 2, cap. 3, qu. 9 art. 2, p. 189.

43. Coimbra Jesuits mention these critiques : *Commenta sunt hominum, qui lucrum ex fallaci admirationum illecebra quaerunt (...)*, De Coelo, liv. 2, cap. 3, qu. 9, art. 3, p. 194.

44. " If by astrology, however, we mean the prediction of natural events and human behaviour on the basis of knowledge of alleged powers inherent in individual celestial bodies and their positions in the heavens, as well as their manifold configurations and interrelationships, then astrology plays very little role in scholastic natural philosophy ", E. Grant, *op. cit.*, 569.

ent astrological practices evinced by Renaissance scholastics like Coimbra
Doctors.

ANÁLISIS DE UN MODELO COSMOGRÁFICO PRESENTADO EN EL COLEGIO IMPERIAL DE LA COMPAÑÍA DE JESÚS EN MADRID

Javier BERGASA LIBERAL

Nos referimos al contenido de un solo folio manuscrito[1] titulado " Conclusiones cosmographicas " numerado como fol. 5R y 5V que está cosido junto con los trabajos, también manuscritos, del Dr. Juan Cedillo Díaz, catedrático de la Academia de matemáticas de la Corte y Cosmógrafo Real entre 1611 y 1625, año de su muerte. Precisamente se trata de algunos de sus tratados astronómicos más interesantes :

- *Dianoya o pensamiento nuebo de los 7 planetas*
- Además hay observaciones del cometa de 1612

La foliación es posterior al encuadernado de los escritos, pues la *Dianoya* que le sigue lleva superpuesta su propia numeración

CONTENIDO DEL MANUSCRITO

Se trata de 31 conclusiones, muy breves y que describen las características del Sistema del Universo. Su contenido está más próximo a los de la Cosmographia o Tratado de Sphera que a las teóricas de los planetas, pues no hay alusiones a los movimientos planetarios ni se intenta describirlos geométrica o físicamente. En consecuencia están más relacionados con la Filosofía natural que con la Astronomía propiamente dicha.

Anotaremos las conclusiones que destacan por su mayor interés astronómico :

1. La Cosmographia es habito del entendimiento contemplativo y descriptivo segun el sitio grandeza movimiento quietud del Universo como finito i compuesto de 4 elementos i cielo empireo.

---

1. Manuscrito de la Biblioteca nacional de Madrid M.S. 9092.

2. El Universo no esta en lugar, antes el es el lugar de todo lo demas, pero su ultima superficie se puede decir estar en lugar i del lugar se estiende cierta distancia desde algun punto determinado como lo es el centro del Universo.

3. La figura del Universo mas conveniente es la espherica aunque della no ai demonstracion, pudiendo tener otras.

4. Por ser el medio i corazon del Universo por su naturaleza se le da a la tierra aunque no siempre el centro de la gravedad sea el mismo que el centro del Universo.

5. Aunque el agua podia tener el centro del Universo pero puesto que la tierra es absolutamente grave puede el agua ceñir la tierra y nadar sobre ella i tambien el aire puede moverse al centro aunque pide estar sobre el agua.

6. El fuego elementar absolutamente es leve no admitiendo gravedad como ni la tierra levedad el agua i el aire son graves i leves respectivamente.

...

22. Unico es el cielo astrifero no de materia dura, sino permeable a los planetas y estrellas.

23. El cielo estrellado es de naturaleza del ayre continuado desde la superficie de la tierra i agua hasta el concavo del fuego que esta contiguo al cielo empyreo.

24. Todas las estrellas i planetas son de materia del fuego elementar i se mueven por ministerio Angelico.

25. Todos los cometas estan en la region de las estrellas sobre la luna y son de materia de fuego ni se hacen de nuebo, ni se corrompen lo mismo siento de las nuebas estrellas que se an aparecido en dias i tiempos entre las demas.

26. En la region Syderea no se admite agua alguna, sino en la Aerea debaxo de la luna donde ban i bienen las nubes.

27. La tierra i agua componen una esphera physica comparativamente pero no matematicamente y tienen un mismo centro de gravedad y sensiblemente es el mismo con el centro de la magnitud pero no matematicamente.

28. Aunque la tierra pueda moverse circularmente de hecho no lo hace ni por su movimiento se pueden salvar todas las apariencias de las estrellas comodamente.

29. Por el bien del Universo el agua i la tierra se mezclan aunque devia el agua rodear la tierra i el fuego en parte se mezcla con el ayre, el qual es rodeado del cielo empyreo fuera del qual no se conoce otro cuerpo.

30. El mundo es uno solo i finito aunque puede aver otros muchos.

31. Las quantidades elementares pueden ser determinadas unas mexores que otras, como lo son la del agua i tierra, sol i luna, Mercurio, Venus y Marte no asi los demas planetas i estrellas ni el elemento del ayre i fuego ni el cielo empyreo.

Las dos primeras conclusiones tratan de forma general de la Cosmografía y el Universo.

Sorprende la posibilidad de una forma no esférica para el Universo en la tercera, ¿se hace eco esta afirmación de las órbitas keplerianas?

Las conclusiones que van de la 6ª a 21ª se detienen en la descripición de las características de los elementos y al movimiento de graves.

En 23 se niega la quinta esencia aristotélica, pues sitúa los planetas en el aire.

En 24 también se niega dicho quinto elemento, pues los planetas son de fuego.

Desaparece del modelo la existencia real de orbes que arrastran a las estrellas y planetas, puesto que se entendía que orbes y estrellas eran de las misma materia con la única diferencia de que la estrella era un especial adensamiento, lo que la hacía brillante y visible, con la imagen conocida de " un nudo e la madera ". Estas conclusiones colocan el modelo descrito en la línea de los que, negando la existencia de los orbes celestes, afirman que planetas y estrellas se mueven como " las aves en el aire o los peces en el agua ".

En 25 se resuelve una de las cuestiones más controvertidas desde el cometa de 1572, su carácter supralunar. El autor así lo defiende insistiendo en su alejamiento del modelo aristótelico.

En 26 se niega la existencia de la esfera llamada Cristalino o aqueo, colocada debajo del primer móvil. Esta décima esfera se usaba en los modelos de 11 esferas (como el de Peurbach, Clavio o Magini) para señalar los elementos fijos de las esfera celeste : polos fijos, eclíptica fija, ecuador fijo, que después se verían sometidos a las diferentes libraciones de la novena y octava esferas.

En 27 trata de la esfera terrestre y resuelve la situación sobre la relación de estos dos elementos : agua y tierra que forma una esfera no matemática, pero sí física, con centro de gravedad.

En 28 apuesta por el modelo geocéntico, aunque considera posible que la Tierra se mueva ya en rotación, ya en traslación.

En 29 vuelve sobre los elementos : Tierra y Agua se mezclan para formar el globo terráqueo, aire i fuego se mezclan pues los planetas y las estrellas, de fuego, se mueven en el aire. Elemento que se extiende desde el globo terrestre hasta el límite del Universo.

En 30 defiende que sólo hay un Universo Mundo, pero parece que esta afirmación procede de describir la realidad conocida, pero no rechaza la posibilidad de que nuevos descubrimientos obliguen a considerar la existencia de otros. Parece que el autor recoge una corriente que tuvo defensores tan destacados como Nicolás de Cusa y muy posiblemente trata de no caer en errores como el de la negación de los habitantes de las antípodas, refutado por vía de la constatación tras los viajes al Nuevo Mundo, o el caso de los cometas, cuyo carácter supralunar fue demostrado por la observación de su paralaje.

En 31 se refiere a las medidas en el Universo. Considera que algunas son sencillas de medir y por la tanto admite como buenos los resultados obtenidos a partir de ellas, por ejemplo el tamaño de la Tierra, la cantidad de agua en el el globo terrestre y su profundidad. Igualmente acepta los valores medidos para el tamaño y distancia de la Luna, Mercurio, Venus y el Sol. No así de los otros planetas. Tampoco la distancia a la que están las estrellas fijas ni sus tamaños reales, en consecuencia tampoco el tamaño de Universo ni la del aire que llena la región desde la Tierra hasta el empíreo.

Se rompe así con la tradición del modelo de Ptolomeo que resolvía ya lo relativo a tamaños y medidas y que a través de Al-fargani llegan hasta los modelos más difundidos durante el siglo XVI.

RELACIÓN DEL MODELO DESCRITO CON LOS PROPUESTOS POR OTROS AUTORES ESPAÑOLES DE LA ÉPOCA

El modelo cosmográfico que se va describiendo en las conclusiones se hace eco de algunas novedades que paulatinamente y respecto de los sistemas tradicionales fueron introducidas por algunos otros autores españoles de la época :

Rodrigo Zamorano (†1620) Piloto Mayor de la Casa de Contratación de las Indias y Catedrático de Cosmografía. Niega la existencia de un esfera del fuego elemental entre el aire y el orbe de la Luna.

Francisco Vicente Tornamira (1534-1597). Niega la existencia real de las esferas sólidas, de los deferente y epiciclos, relegándolos a un simple instrumento que permite seguir sus movimientos, describiendo y anticipando las posiciones

El movimiento de los cuerpos celestes no es pues el de sus orbes o esferas, sino el de lo mismos cuerpos. Vemos en ello un ejemplo de lo que supone el conocido símil de que las estrellas se mueven libremente como las aves en el cielo o los peces en el agua.

Jerónimo Muñoz (*ca.* 1520-*ca.* 1591). Catedrático de Hebreo y Matemáticas en Valencia y de Astrología en Salamanca. Se muestra partidario de la mutabilidad de los cielos.

Existen referencias indirectas, como la de Pérez de Mesa, que sitúa a Muñoz entre los que niegan la existencia de los orbes celestes e incluso hace referencia a un supuesto tratado de Muñoz sobre la inexistencia de orbes.

Pedro Simón Abril (1530-*ca.* 1595) Profesor de Humanidades en la Universidad de Zaragoza. Manifiesta su desacuerdo con la supuesta inmutabilidad de los cielos y se apoya en los dos " cometas " que el mismo observara, considerando por ello que los cambios no se producen únicamente en la región elemental

Respecto al fuego elemental dice que o puede asegurar como verdad cierta e infalible que lo haya, aunque parece cosa muy conforme a la razón y muy probable

Pérez de Mesa (†*ca.* 1635) Profesor en Alcalá y Catedrático de Matemáticas en Sevilla. Para él no hay esferas, orbes o círculos, puesto que defiende que los planetas se mueven libremente como peces en el agua o pájaros en el cielo. Lo que tanto como decir que sólo hay una esfera : la celeste.

No hay una materia propia de la región celeste, sino que es el mismo aire que envuelve a la esfera terrestre que continúa por la región celestial.

También analiza la posibilidad del movimiento de la tierra y lo encuentra posible, pero termina por creer más probable su quietud.

Las estrellas y planetas no están formadas por un elemento diferente — la quinta esencia de aristotélica — ni tampoco de aire — elemento a través del que se mueven — sino compuestos de diversos elementos como cualquiera de los objetos o cuerpos que hay en la Tierra, sigue en esto a Platón

Juan Cedillo Díaz (†1625) Catedrático de la Academia de Matemáticas de la Corte y Cosmógrafo Mayor. Defiende que los planetas se mueven por el aire recorriendo los círculos deferentes y epiciclos, pero sin que tengan estos existencia real[2].

¿ANTE QUIÉN SE DEFENDIERON LAS CONCLUSIONES?

Constaban los Estudios de 23 cátedras, seis de ellas menores. Entre las mayores existen algunas relacionadas con la enseñanza de temas astronómicos :

" VII De filosofía natural para leer la física, los dos libros de la generación y corrupción, los tres de Coelo y el cuarto de meteoros.

IX. De matemática donde un maestro por la mañana leerá la esfera, astrología, astronomía, astrolabio, perspectiva y pronósticos.

X. De matemática donde otro maestro diferente leerá por la tarde la geometría, geografía, hidrografía y de relojes ".

Las primeras lecciones de 1629 fueron impartidas por el P. Francisco Ruiz en la cátedra de Física y por el P. Juan Bautista de Poza las otras dos, por estar vacantes dichas cátedras, y además dio la lección de Placitis philosophorum, que era su cátedra[3].

Vemos, pues, que el modelo presentado y discutido en los Reales Estudios del Colegio Imperial supone romper con los modelos tradicionales y tiene un referente claro y próximo en las obras de bastantes de los astrónomos y cosmógrafos españoles que trabajaron y escribieron sus tratados a finales del siglo XVI y principio del XVII. Lo que supone al autor y a los PP. Jesuitas bien ente-

---

2. Los datos que se asocian con cada autor proceden de las siguientes obras : R. Zamorano, *Compendio de la arte de navegar*, Alonso de la Barrera, Sevilla, 1581 ; F.V. Tornamira, *Chronographia y repertorio de los tiempos*, Tomás Porralis, Pamplona, 1580 ; J. Muñoz, *Libro del nuevo Cometa*., Pedro de Huete, Valencia, 1573 ; P. Simón Abril. *Philosophia Natural* . s.f. Manuscrito ; D. Pérez de Mesa, *Comentarios de Sphera*. Manuscrito (1596) ; J. Cedillo. *Ydea cosmographica o Fabrica del Mundo*, s.f., Manuscrito.

3. J. Simón Diaz, *Historia del Colegio Imperial de Madrid*, Madrid, CSIC, 1952, vol. I.

rados no sólo de las publicaciones de los grandes astrónomos europeos del momento, sino también de cuanto se hacía en España.

Sin embargo, aunque no conocemos en profundidad qué se enseñó en estos Estudios Reales en us primeros cursos, sí tenemos el testimonio de Lope de Vega, el famoso dramaturgo español, que estudiara en el primer colegio de los jesuitas y también en la Academia de Matemáticas de la Corte. Quien en una obra en verso, *Isagoge a los Reales Estudios de la compañía de Jesús*, encargada por el propio rector para celebrar el comienzo de curso y publicada en 1629 va loando a lo largo de los 720 versos que la forman a los protectores de los estudios y a sus profesores, mientras glosa los contenidos de la primera lección de cada materia. Nada parece indicar que en lo referente a la Astronomía, el P. Juan Bautista Poza encargado de a primera lección tratar de contenidos diferentes de los tradicionales[4].

DATACIÓN Y AUTORÍA

Una dificultad este folio 5 no aparece datado, pero podemos situarlo con la información siguiente.

El Colegio de la Compañía de Jesús atraviesa tres épocas :

1. Casa y Colegio (1560-1602)
2. Colegio Imperial (1603-1625)
3. Los Reales Estudios (1625-1767)

Debe ser posterior a Enero de 1625 por que en esa fecha se da la Real Cédula de fundación de los Estudios Reales indicándose ya las 23 cátedras que las componían, aunque finalmente serían sólo 22 al desaparecer la de Cánones por presión de las Universidades de Salamanca y Alcalá[5].

Podemos establecer dos hipótesis de trabajo :

Hipótesis A.

Si se tratase de Febrero de 1625 las conclusiones serían sin duda de Cedillo, quien se habría reunido con los padres jesuitas encargados de esas cátedras para tratar de los contenidos a explicar, instándoles a incorporara las más modernas ideas. Lo que nos situaría ante un documento de capital importancia para conocer la configuración que da al Universo el astrónomo español más destacado de la época.

Además la relación entre Cedillo y los PP. Jesuitas parece evidente toda vez que a su muerte ellos asumen la enseñanza en la Academia de Matemáticas de la corte durante los cursos 1625-26, 26-27 y 27-28 y después la Cátedra de Matemáticas de la Corte pasa a los Estudios Reales del Colegio la veracidad de esta situación queda satisfecha por la existencia de una real de Cédula de

---

4. *Idem.*
5. *Idem.*

Felipe III, ordenando " que se pague al Colegio Imperial el salario pendiente por la lectura de las matemáticas desde el día de la muerte de Cedillo hasta el día de San Lucas de 1628 " (S. Lucas es el 18 de Octubre y las clases comenzaban precisamente el siguiente día 19 de Octubre)[6].

Se podrían plantear dos dificultades a esta hipótesis A que atribuye a Cedillo la autoría :

1. Los contenidos de las conclusiones parecen de acuerdo con los últimos trabajos de Cedillo, sobre todo con la *Ydea Cosmographica* que es una traducción parcial de *De Revolutionibus* de Copérnico, los dos primeros libros completos y parte del tercero. Sin embargo no hay constancia de que él enseñara el modelo copernicano en sus clases de la Academia, ¿no sería muy osado por su parte instar a los padres Jesuitas para que sí lo hicieran?

2. Si Cedillo asume el modelo copernicano, ¿por qué en la conclusión 28 niega el movimiento terrestre?

A la primera dificultad se podría objetar que no existe mención de sus enseñanzas ni sobre quiénes o cuántos eran sus alumnos.

A la segunda se puede objetar que o bien él mismo había rectificado y cambiado el modelo heliocéntrico por otro, como pueda ser el de Tycho que también conocía, o bien pensó que los padres no aceptarían el modelo original copernicano tras la condena de Galileo y presentó uno que sí pudieran considerar como aceptable.

Hipótesis B.

Si el autor fuera posterior muy bien podría tratarse de alguno de los jesuitas del Colegio quien planteara a discusión este modelo, lo que nos situaría ante alguien bien preparado en Sphera y teórica de los planetas (es decir Astronomía), lo que parece contradecir las dificultades que tuvieron para encontrar una persona que se hiciese cargo de la enseñanza de las Astronomía hasta que llegó el P. Claudio Ricardo (1589-1664) de paso hacia las Indias y a quien se obligó a quedarse para que se hiciera cargo de la cátedra IX de los Estudios y a quien se concedieron los títulos de Catedrático de Matemáticas de la Corte y Cosmógrafo Real a partir de Agosto de 1629[7].

No se conoce trabajo alguno de Claudio Ricardo en España, aunque sí aparece como autor de un informe sobre una Aritmética y como firmante de la aprobación a una traducción de los *Elementos de Euclides*[8].

6. V. Maroto, M.I. y E. Piñeiro, *Aspectos de la Ciencia aplicada en la España del siglo de Oro*, Salamanca, Junta de Castilla y León, 1991.

7. *Idem*.

8. J. Simón Diaz, *Jesuitas de los siglos XVI y XVII : escritos localizados*, Madrid, Universidad Pontificia de Salamanca, 1975.

Conclusiones

Incluso si su autor fuese cualquier estudiante del Colegio — extraño en tanto que la institución no estaba capacitada para emitir títulos, por lo que no puede tratarse de un trabajo para la consecución de grados — estaríamos ante un texto verdaderamente importante dentro de la evolución de la astronomía española de la época, pues el empuje y vitalidad demostradas en la segunda mitad del S. XVI y principios del XVII van decayendo tras la muerte de Cedillo y no hay referencia de autores o escritos que se aparten del modelo de Peurbach o similares, siendo el de Tycho el más avanzado de los que se utilizan[9], al extremo que todavía en 1748 — 100 después de los textos y autores estudiados aquí — Jorge Juan se ve obligado a tildar de hipotéticas y supuestas las teorías heliocéntricas y newtonianas en la introducción de su *Observaciones astronómicas y phisicas*[10].

Por otra parte, rechazar estas conclusiones, u otros textos que rompieran con la tradición ptolomáica, ya de la voz de Cedillo o de cualquier otro y mantener una enseñanza de las astronomías basada en la obra de Clavio — *In Spheram de Ioannis de Sacrobosco commentarius*, Roma, 1581 — o en las de Conrado Dasypodio — *Heron mechanicus*, 1580, *Institutionum Mathematicarum*, 1593, o *Elementorum arithmeticae, geometriae, opticae, catoptricae, scenographiae, theoricae planetarum*, 1594 — que sí se utilizaban en el Colegio, fue causa de un importante retraso en la difusión de ideas y modelos cosmográficos más modernos. Sobre todo, si se tiene en cuenta que por encontrarse en la Corte fue, el Colegio Imperial, el centro de enseñanza al que acudían la inmensa mayoría de los jóvenes que acabarían teniendo todo tipo de responsabilidades en la segunda mitad del siglo XVII.

9. Un buen ejemplo lo encontramos en Tomás Vicente Tosca, *Compendio Mathematico*, Valencia, A. Bordázar, 1707-15, vol. I-IX.

10. J. Juan y Santacilia y A. de Ulloa,*Observaciones astronómicas y phisicas hechas por orden de su Magestad en los Reynos de Perú por...de los quales se deduce la figura y magnitud de la Tierra,* Madrid, Juan de Zuñiga, 1748. Es muy intersante comparar la introducción de esta primera edición con la de 1773, editada en la Imprenta Real de la Gazeta, para observar las sutiles diferencias al tratar del sistema heliocéntrico y de la teoría newtoniana y conocer la dificultad en la que todavía se hallaban el estudio y difusión de esas ideas, aunque la de Copérnico tuviese ya dos siglos de difusión. Más palpable se nos muestra la penuria del desarrollo de todas estas ideas si leemos la obra de J. Juan, *Estado de la astronomía en Europa y juicio de los fundamentos sobre los que se erigieron los sistemas del mundo...*, Madrid, Imp.real de la gazeta, 1774. Obra póstuma en la que el autor se duele de que todas las teorías no geocéntricas deban ser acompañadas de " no se crea éste que es contra las Sagradas Letras " (*cf.* pág. 15).

ENTRE SCIENCE ET PHILOSOPHIE : LA COSMOLOGIE DANS L'*ENCYCLOPÉDIE*.

D'ALEMBERT ET MAUPERTUIS

Martine GROULT

Dans le respect que d'Alembert porte à Descartes, ce qui est à retenir c'est l'affirmation inébranlable et maintes fois répétée qu'il ne faut en aucun cas confondre " sa cause avec celle de ses sectateurs " (*DP*, xxjx)[1]. Cela signifie, pour d'Alembert, que le progrès des connaissances humaines ne revient qu'aux " génies ". Deux qualités leur sont spécifiques, une qui relève de la conduite : avoir une démarche courageuse, et une qui relève des fondements de la science : avoir énoncé les principes généraux de la science. D'Alembert a envers Aristote[2] et envers Leibniz[3] la même appréciation qu'envers l'auteur de la *Dioptrique*. Dans l'*Encyclopédie*, il porte le même jugement sur Maupertuis. Il lui attribue les deux qualités du génie, premièrement le courage : " Le premier qui ait osé parmi nous se déclarer ouvertement newtonien, est l'auteur du *Discours sur la figure des astres* " (xxjx) [1732] et deuxièmement, l'énoncé d'un principe général : le principe de moindre action. La seconde considération est, pour tout savant, le résultat d'une démarche scientifique. D'Alembert nomme cette démarche *sagacité*. Elle désigne l'esprit observateur[4]. L'approche

---

1. L'abréviation *DP* désigne le Discours préliminaire de l'*Encyclopédie ou dictionnaire raisonné des sciences, des arts et des métiers* de Diderot et d'Alembert (1751-1765 pour les volumes de discours et 1762-1772 pour les volumes de planches). Nous plaçons " art. " devant les noms des articles de l'*Encyclopédie*. Tous les articles cités sont de d'Alembert.

2. Ne mettons pas, écrit d'Alembert à l'art. *Expérimental,* sur le compte d'Aristote " l'abus que les modernes en ont fait, durant les siècles d'ignorance qui ont duré si longtemps, ni toutes les inepties que ses commentateurs ont voulu faire prendre pour les opinions de ce grand homme ", (1756), t. VI, 299a.

3. Lettre à Frédéric du 3 Novembre 1764 où Descartes et Leibniz sont considérés l'un comme l'autre " homme de génie " par d'Alembert qui reproche au Roi de Prusse de leur préférer Bayle et Gassendi.

4. D'Alembert considère Diderot comme le philosophe qui porte " dans l'étude de la nature la sagacité & la sagesse de l'esprit observateur ", art. *Elasticité* ou *Force Elastique*, (1755), t. V, 445a.

du savant se limite d'abord à l'observation de la nature et demande de la sagacité. Et, à l'époque, il fallait du courage pour partir des choses de la nature et
non de Dieu, ce qui explique la nécessité de posséder la première qualité.

Pour énoncer son principe, Maupertuis est parti de l'observation suivante :
la nature ménage dans tout changement une quantité d'action minimum (art.
Cosmologie, t. IV (1754), 295b). L'auteur de l'*Essai de cosmologie* présente
ainsi sa démarche : " je ne me suis attaché qu'aux premiers principes de la
Nature, qu'à ces Loix que nous voyons si constamment observées dans tous les
Phénomènes "[5].

L'observation apporte à la découverte d'un principe, la qualité cruciale de
l'invariance sans laquelle le principe ne peut être un principe général de la
science.

Ce point de vue nouveau porté par d'Alembert sur les savants entraîne deux
conséquences. La première est une rupture entre ceux qui font la science et les
" sectateurs " qui l'expliquent. La seconde est un nouveau rapprochement entre
la science et la philosophie. Voyons ces deux points.

LA RUPTURE ENTRE SAVANTS ET SECTATEURS

La rupture entre savants et sectateurs est aussi une rupture entre la science
et un certain type de philosophie : celle qui prend les mots pour les choses.
Maupertuis est d'abord un géomètre qui apprécie dans la géométrie le fait que
ce soit une science qui dise " comment les choses sont ". L'Avertissement de
1750 de l'*Essai de cosmologie* décrit ce fait. Dans la première édition, Maupertuis est plus occupé de présenter sa démarche pour expliquer le système du
monde, que de se défendre contre ses adversaires, comme dans les deux éditions suivantes de 1752 et 1756[6]. Il décrit ce qu'il a fait et cela ne sera plus
reporté dans les versions ultérieures. Dans les deux parties de l'ouvrage,
l'auteur insiste sur la manière dont il a présenté la première partie. Il s'agit, dit-
il, d'" une exposition du Système du monde, mais que je ne présente que
comme un tableau, & non comme une explication ". Dans cette première partie, " j'ai tâché, de pénétrer jusqu'aux premières raisons sur lesquelles les loix
de la Nature étoient fondées : dans l'autre j'ai tâché de bien peindre les phénomènes de l'Univers ".

---

5. G. Tonelli, " Maupertuis et la critique de la métaphysique ", *Actes de la Journée Maupertuis*,
Paris, Vrin, 1975, 87.
6. Trois éditions existent de l'*Essai de cosmologie*. La première édition de 1750 (Berlin, cote
BN : V 11968) comporte un Avertissement, la seconde édition de 1752 (*Œuvres* de Mr. de Maupertuis, Dresde, G.C. Walther) contient une Préface et la troisième de 1756, un Avant-propos. La
différence la plus conséquente est entre l'Avertissement de 1750 et la Préface de 1752. L'Avant-
propos de 1756 (réédition posthume sans changement en 1768), est reproduit dans l'édition de
G. Tonelli, *op. cit.*, t. IV, 3-28.

La volonté de mettre en tableau avant d'expliquer veut dire que la priorité est donnée aux choses et non à ce qui les explique, à savoir les mots. Dans son éloge prononcé le 24 janvier 1760, Formey — secrétaire perpétuel de l'Académie des Sciences et Belles Lettres de Prusse — rapporte que, digne académicien, Maupertuis a fait ce que les académiciens des sciences ont coutume de faire : " mettre les choses à la place des mots ". Lorsqu'après avoir évoqué les temps ténébreux pour la science, d'Alembert fait l'apologie de " ces quelques génies supérieurs, qui abandonnant cette méthode vague & obscure de philosopher, laissoient les mots pour les choses, & cherchoient dans leur sagacité & dans l'étude de la Nature des connoissances plus réelles " (art. *Expérimental*, t. VI (1756), 299a) il affirme la même rupture. Cette sagacité appartient aux savants qui ont l'esprit observateur et non à leurs sectateurs qui n'ont pas pris l'étude de la nature pour unique point de départ. Les sectateurs ne font que des dissertations pendant que les savants participent à la découverte des maillons de la chaîne des connaissances que sont les principes généraux de chaque science. Le premier à qui revient le mérite d'avoir mis en avant le point de vue du savant est Francis Bacon[7]. Depuis le chancelier anglais, le tableau des lois qui gouvernent le système du monde consiste dans le tableau des choses. Il est ouvert à toutes les possibilités d'arrangements à partir du moment où les choses qui y sont disposées relèvent de ce qui a été observé dans la nature. Avant qu'il soit question d'ordonnancement, il est question d'observation, ce qui conduit à poser des limites et à former l'esprit selon une certaine rigueur. Les sectateurs qui disputent sur les fondements religieux des théories scientifiques, établissent un mélange entre principe métaphysique et principe mathématique. Ils sont à l'origine des querelles au lieu d'être à l'origine des découvertes.

D'Alembert apprécie chez Maupertuis sa précision rigoureuse sans référence à une efficace qu'elle recevrait de Dieu[8]. Maupertuis règne, écrit Gueroult, dans un univers " d'où est exclu tout absolu se conservant "[9]. D'Alembert cite dans l'article de l'*Encyclopédie* la préface de l'*Essai de cosmologie* de 1752 qui, nous allons le voir, privilégie le regard sur les choses, et donc les expériences, pour placer Dieu à l'extérieur du fonctionnement des corps.

7. On peut se reporter à B. Milner, " Francis Bacon : The theological foundations of Valerius Terminus ", *Journal of the History of Ideas*, 58, 2 (1997), 256-262. L'auteur souligne que c'est chez Bacon qu'apparaît pour la première fois la distinction entre la croyance ou la foi et la connaissance.

8. La Lettre XVII *Sur la Religion*, Maupertuis écrit : " La Religion ne tient point aux autres parties de nos connaissances ; elle n'est appuyée, ni sur les principes des Mathématiques, ni sur ceux de la Philosophie : ses dogmes sont d'un ordre qui n'a aucune liaison avec aucun autre ordre de nos idées, & forment dans notre esprit une Science entièrement à part, qu'on ne sauroit dire qui s'accorde ni qui répugne avec nos autres Sciences ". dans *Œuvres complètes*, ed. Tonelli, *op. cit.*, t. II, 327.

9. M. Gueroult, " Note sur le principe de la moindre action chez Maupertuis ", *Leibniz, Dynamique et Métaphysique*, Paris, Aubier, 1967, 233-234.

On est ainsi arrivé à une nouvelle séparation : non plus celle entre savants et sectateurs, mais celle entre les choses et les mots, c'est-à-dire entre la science et la philosophie de l'époque menée par ceux qui ne savent que discuter, voire disputer, à partir de points de vue religieux. L'*Essai de cosmologie* fonde cette séparation pour installer le nouveau point de départ de l'ordre des connaissances. On a pu justement parler de retournement des perspectives[10] effectué par Maupertuis. D'Alembert conclut que le principe du minimum dégagé par Maupertuis restera toujours " une vérité géométrique " (art. *Action*, t. I (1751), 120b) et qu'il " n'est donc, comme tous les autres, qu'un principe mathématique " (art. *Cosmologie*, 297b) et non un principe qui concilie métaphysique et mathématique, ainsi qu'il l'avait annoncé à l'article Action (119a-120a). Mais attention, si ce n'était que cela, c'est-à-dire la mise à l'écart de Dieu ou d'une certaine métaphysique, il n'y aurait aucune ouverture et d'Alembert ne serait guère intéressant, pas plus que Bacon. L'expérience ne veut pas dire les choses. Elle veut dire le rapport entre les choses. Il s'agit alors moins de supprimer Dieu que de faire intervenir l'esprit de l'homme. La perception des rapports énoncée dans l'*Encyclopédie*, veut dire la valorisation du travail effectué par l'esprit humain.

LE NOUVEAU RAPPROCHEMENT ENTRE SCIENCE ET PHILOSOPHIE

a) Le retournement des perspectives

Le point de départ n'est pas les idées de Dieu en nous, mais l'observation des phénomènes qui nous montre ce que Dieu a fait. Ce retournement des perspectives réalisé par Maupertuis est exposé par d'Alembert à l'article Causes finales (t. II (1751), 789b) : " s'il est dangereux de se servir des causes finales à priori pour trouver les lois des phénomènes ; il peut être utile, & il est au moins curieux de faire voir comment le principe des causes finales s'accorde avec les lois des phénomènes, pourvû qu'on ait commencé par déterminer ces lois d'après des principes de méchanique clairs & incontestables. C'est ce que M. de Maupertuis s'est proposé de faire à l'égard de la réfraction en particulier, dans un mémoire imprimé parmi ceux de l'Académie des Sciences, 1744 ".

C'est donc bien par les principes observés dans la nature et démontrés par l'expérience qu'il faut commencer pour découvrir les phénomènes. Aux premières personnes qui ont présenté l'*Essai* comme une œuvre qui étouffe les preuves de l'existence de Dieu, Maupertuis répond en 1750[11] : " ...ce que j'ai dit c'est qu'on a trop multiplié ces preuves, qu'on a quelquefois pris pour des preuves ce qui n'en étoit point ; & que je conseille de chercher la Démonstra-

---

10. J. Roger souligne que Maupertuis rompt avec un esprit scientifique vieux de trois quarts de siècle, *Les Sciences de la vie dans la pensée française au XVIIIe siècle*, Paris, 1963 (rééd. A. Michel, 1993), 481-483.

11. L'Avertissement de 1750 ainsi que la première partie de l'*Essai* avaient été imprimés dans le tome II des *Mémoires de l'Académie royale des Sciences et Belles Lettres de Prusse*.

tion de l'Existence de l'Etre suprème dans les Phénomènes universels de la Nature plûtôt que dans ses petits détails ".

La conséquence du renversement consiste à placer Dieu ailleurs que dans la nature observable par ses " petits détails ". Pareille conséquence n'échappera pas à Condillac qui écrit à Maupertuis le 10 juin 1750 : " Il me paraît hors de doute que ce n'est pas dans les petits détails qu'il faut chercher l'être suprême, le grand nombre d'exceptions auxquelles ils sont sujets sont inextricables (…). Je suis charmé de me voir confirmé dans mes idées en me rencontrant avec vous… ". L'universel est le domaine de Dieu, les petits détails celui de l'homme. La séparation est consommée. En 1752, alors qu'il fait une plus large place aux critiques, Maupertuis s'adresse aux pieux " Contemplateurs " de la Nature (déistes autant qu'athées) : " J'ai dit que ce n'étoit point par ces petits détails de la construction d'une plante ou d'un insecte[12], par ces parties détachées dont nous ne voyons point assez le rapport avec le Tout, qu'il falloit prouver la puissance & la sagesse du Créateur : que c'étoit par des Phénomènes dont la simplicité & l'universalité ne souffrent aucune exception ".

Dans la seconde édition de l'*Essai*, les " petits détails " ont pris de l'importance. Le but maupertuisien est de faire du rapport entre le Tout et Dieu, un domaine à part : le domaine de la contemplation différent de celui de l'observation. En 1756, dans la troisième édition, il ajoute à la fin de la phrase ci-dessus citée : " & ne laissent aucun équivoque ". Les " Phénomènes " qui prouvent Dieu ne relèvent pas des expériences qui prouvent les lois des phénomènes. Dieu est dans l'Universel, mais l'homme doit passer par les " petits détails " de l'univers pour comprendre l'Universel ou, pour parler en termes newtoniens, pour comprendre le système du monde. L'universel de l'homme est le résultat de la rigueur et de la précision scientifique que seule l'observation et la description strictes des détails permettent d'acquérir.

Nous nous trouvons devant un nouveau point de vue : celui de l'homme seul face aux choses. Il a alors un tableau à expliquer. Remarquons que ce tableau, c'est lui qui l'a établi. A lui donc de l'ordonner. Le Tout possible à voir est celui qui est à construire avec les facultés de l'esprit humain. L'homme entre dans l'universel non pas en essayant de se mettre à la place de Dieu, mais en analysant le travail réalisé par son esprit. Tel est ce que d'Alembert explique en 1755 au début de l'article Eléments des Sciences de l'*Encyclopédie*. C'est uniquement par les liaisons que l'esprit du *génie* découvre et par suite, que l'homme pourra comprendre les phénomènes inscrits sur son tableau.

b) La cosmologie est une totalité logique

Le point de départ étant l'observation et l'expérience, l'homme est devant les choses précises dont il peut établir des définitions. Pour passer de la saga-

---

12. Maupertuis désigne Malebranche, *De la Recherche de la vérité*, Livre I, ch. VI.

cité à l'universalisation, l'intervention du philosophe sera nécessaire. Cela veut dire, dans les termes dalembertiens : passer de " l'esprit observateur à l'esprit systématique ", plus communément appelé postérieurement, " l'esprit philosophique ". C'est ce que démontre l'article Cosmologie. Avant d'affirmer la totalité, c'est-à-dire que tout est lié dans la nature, cet article définit étymologiquement le terme, qui en grec veut dire " monde " et " discours " : " Ce mot (…) signifie à la lettre science qui discourt sur le monde, c'est-à-dire qui raisonne sur cet univers que nous habitons (…). La Cosmologie est donc proprement une Physique générale & raisonnée (…) " (art. *Cosmologie*, 294a).

La totalité n'est plus maintenant le Tout à contempler, mais un discours à analyser. Les savants qui pratiquent la philosophie naturelle — Maupertuis se défendra d'être un métaphysicien pour affirmer qu'il est un naturaliste — au début du XVIIIᵉ siècle s'attachent à démontrer qu'il y a une logique, un *logos* possible. La logique en tant que " discours scientifique sur " est la pratique du raisonnement qui accède aux principes généraux d'une science. La cosmologie est la science qui a pour objet la compréhension générale de l'Univers, c'est-à-dire qu'il s'agit d'une science qui explique les faits par les principes généraux. Pour d'Alembert, la cosmologie est une physique générale et raisonnée : cela signifie qu'elle est une physique métaphysique par rapport à une physique uniquement expérimentale. C'est pourquoi il la distingue de la cosmographie et de la cosmogonie.

La cosmographie est une description du monde : " science qui enseigne la construction, la figure, la disposition, & le rapport de toutes les parties qui composent l'Univers " (art. *Cosmographie*, 293a-b). La cosmogonie " est la science de la formation de l'Univers " (292b). Donc la cosmologie n'est ni l'histoire de l'univers ni l'enseignement des choses et de leur rapport. Tout cela constitue la physique expérimentale, laquelle procure les matériaux nécessaires à la physique générale et raisonnée. Fontenelle, dans la *Préface sur l'utilité des Mathématiques et de la Physique*, dénommait cette autre physique conséquente de l'expérience : la " Physique systématique "[13].

De fait, la cosmologie dispense autre chose que les résultats de l'expérience. Fondée sur eux, elle donne leurs règles et leurs raisons. Elle est métaphysique, c'est-à-dire qu'elle donne le point de vue de l'esprit humain en train de construire le rapport entre les choses. Rien n'est mystérieux ici, car tout est analysable. Entre les choses et les mots, d'Alembert construit un espace plein qui n'est ni la matière ni Dieu, mais le travail de l'esprit humain. Cet espace, décrit dans le *DP* comme un intervalle immense, est le point fondamental de l'épistémologie dalembertienne. A partir de lui, l'auteur du *Traité de dynamique* établit une distinction entre " l'esprit de système et l'esprit systématique ". Non pas la mise en rapport, mais la mécanique de la mise en rapport ; celle qui va

---

13. Fontenelle, *Préface sur l'utilité des Mathématiques et de la Physique et sur les travaux de l'Académie des Sciences*, 1702, (Paris, Fayard, 1994), vol. VI, 49.

permettre la liberté de l'esprit[14]. S'il met en rapport, la plupart du temps, le savant n'explique pas le mécanisme mis en œuvre par son esprit pour arriver à la découverte. Or, ce mécanisme est celui de la vérité que le philosophe découvre au savant.

Le cartésianisme de d'Alembert est situé sur ce point précis où la méthode se révèle être le seul outil qui délivre la démarche de l'esprit : on ne peut se passer d'une méthode pour se mettre en quête de la vérité des choses (Descartes, Règle IV des *Règles pour la direction de l'esprit*). La méthode est le domaine de l'explication et non plus du tableau. Elle est méthode générale pour expliquer toute matière, à condition que les expériences nécessaires aient été prises en considération (Descartes, Lettre LXXIV du 27 avril 1637). Notons l'exigence de situer la méthode selon l'expérience qui discriminera les explications et fera de la méthode une méthode générale. Le savant alors quitte son " esprit de système " et revêt l'" esprit systématique ". Il travaille selon une logique dans laquelle l'universel apparaît au point de sensibilité de l'application d'une science à une autre science. L'" esprit systématique " décrit la marche à suivre par l'esprit, à savoir la réduction : " plus on diminue le nombre des principes d'une science, plus on leur donne d'étendue ; puisque l'objet d'une science étant nécessairement déterminé, les principes appliqués à cet objet seront d'autant plus féconds qu'ils seront en plus petit nombre. Cette réduction, qui les rend d'ailleurs plus faciles à saisir, constitue le véritable esprit systématique… " (*DP*, vj).

Cette réduction représente le travail réalisé par le savant lorsqu'au cours d'une invention ce dernier opère le passage des détails aux principes généraux applicables à une autre science que la sienne particulière. Il consiste dans l'esprit qui raisonne à partir des choses. Le point de vue est ici celui de l'abstraction. Le travail de l'esprit désigne l'entendement humain en train de réaliser les liaisons, à partir de l'expérience, entre les propositions de la science particulière.

Tel est l'aspect ingénieux de l'épistémologie de d'Alembert qui, partant de sa propre découverte (le principe de l'équilibre), de celle de Maupertuis (le principe de moindre action) et de l'invention faite en Angleterre par Newton (le principe de l'attraction rapporté en France par Maupertuis et par Voltaire), fournit une explication métaphysique à ces découvertes. Il s'agit d'une nouvelle approche cosmologique pour expliquer le système du monde. Dans l'article Cosmologie (294a), l'Universel intervient après les définitions ci-dessus rappelées avec la phrase " tout est lié dans la Nature ". Après l'observation des savants, l'art du philosophe (294a) sera de lier toutes les démarches des savants. Le retournement des perspectives chez Maupertuis est confirmé chez d'Alembert dans un retournement de la conception de la chaîne des connais-

---

14. Sur ce sujet que nous ne pouvons pas développer ici, mais qui a constitué l'objet de notre thèse, voir Adorno, *Dialectique négative*, Paris, Payot, 1978, 27-32.

sances. Elle est celle de l'entendement humain et non celle de Dieu que l'homme chercherait à comprendre. Lier les démarches des savants ne veut pas dire lier tous les phénomènes du spectacle de la nature. Dans l'*Encyclopédie*, comme chez Maupertuis, l'homme n'est pas un spectateur ou un " contemplateur " de la Nature comme le donnait l'abbé Pluche : il est un observateur. Le créateur des maillons et de la chaîne, c'est l'esprit de l'homme. L'entendement humain ajoute " de nouveaux chaînons aux parties séparées ". Le retournement de perspective conduit au tableau ou " Système figuré des connaissances humaines " dans lequel la représentation de l'Univers consiste dans l'entendement humain qui a observé la nature. L'*Encyclopédie* nous en livre le " discours " de l'entendement.

La matière des chaînons sont les faits et leurs liaisons sont les lois générales (art. *Cosmologie*, 294a). Les lois générales s'observent dans un grand nombre de phénomènes, mais pas dans tous. En effet, d'Alembert souligne : " je me garde bien de dire dans tous ". C'est parce qu'il y a des lois générales — découvertes à partir des petits détails — qu'il n'y a pas de loi unique non analysable : " nous ne connaissons pas tous les faits, comment pourrions-nous donc assurer qu'ils s'expliqueront tous par une seule et unique loi ? " (294b).

L'universalisation n'est surtout pas systématisation mathématique[15] ce qui aurait l'air de supposer qu'une science comprendrait Dieu. Il n'y aurait, en ce cas, aucun retournement des perspectives. Maupertuis se contente d'exposer un système de l'Univers, ou selon sa propre expression, un " tableau ", parce que c'est tout ce que l'homme peut faire (295a). Pour l'homme, seule la multiplicité des principes conduit à la vérité. Le scepticisme est bien plus source de progrès pour le savoir qu'on a pu le dire jusqu'à présent. Le fait de partir de l'observation et d'analyser à partir de ce que l'on voit, ouvre une *route* immense : celle de l'esprit de l'homme au travail. L'homme a un rapport aux choses qui est une représentation de rapports particuliers dont la connexion est possible par son esprit, seul élément capable de totalité. Citons en dernier lieu sur ce point, une lettre de Maupertuis dans laquelle tout lecteur de l'article Eléments des Sciences reconnaît la similitude de pensée. Il s'agit de la Lettre XVIII, *Sur la divination*[16] : " Ce n'est pas que tout étant lié dans la Nature, un esprit assez vaste ne pût, par la petite partie qu'il aperçoit de l'état présent de l'Univers, découvrir tous les états qui l'ont précédé, & tous ceux qui doivent le suivre : mais nos esprits sont bien éloignés de ce degré d'étendue. La mémoire ne nous représente point le passé par la vue de la connexion qu'il a avec le tout : elle ne nous le rappelle que par des rapports particuliers qu'il a avec notre perception présente ".

15. G. Tonelli, *La pensée philosophique de Maupertuis, son milieu et ses sources*, Zürich, New York, Olms Verlag, 1987, 29 sqq.

16. *Œuvres complètes*, ed. Tonelli, *op. cit.*, t. II, 332.

Le savant et le philosophe sont réunis par la perception. On a donc retrouvé un nouveau type de conjonction entre science et philosophie. Nouveau parce que diffèrent de celui qui a été dénoncé chez les sectateurs et qui ne pouvait conduire à une entente réelle entre les deux domaines. La liaison entre science et philosophie consiste dans la voie de la méthode qui permet de rendre compte de la perception des phénomènes. C'est la méthode analytique lorsqu'elle est l'explication par le génie du travail de l'esprit humain.

CONCLUSION

Nous pouvons conclure qu'il est question avec la cosmologie de construire une chaîne dont les chaînons sont les inventions des savants. L'histoire des sciences s'attache à l'histoire de la découverte des principes qui sont le résultat de la démarche de l'esprit du " génie ". Le lien entre les inventions est possible par l'étude des principes généraux de chaque science. Toutefois cela ne saurait constituer une chaîne. Une totalité n'est entreprise qu'avec la méthode utilisée par le savant, c'est-à-dire la démarche suivie par son esprit. Or, ce dernier point est le sujet de la philosophie que l'*Encyclopédie* définit strictement comme l'art de comparer, autrement dit de mettre en rapport. La seule affirmation dalembertienne — et Maupertuis la partageait avec lui — n'est ni celle d'une loi unique, encore moins celle d'une méthode unique, mais celle d'un ordre. Mettre en rapport, c'est savoir mettre en ordre. Il y a un ordre, arbitraire entre les sciences, mais originaire de la méthode que le savant emploie et que le philosophe explique. L'ordre est le révélateur de la logique du " génie inventeur ". La cosmologie est la conception de la chaîne du monde par l'esprit humain, c'est-à-dire par cette logique que dévoile l'histoire des sciences.

En tant que " discours sur le monde ", il revenait à la cosmologie de montrer que l'histoire des sciences n'est pas l'histoire des faits scientifiques, mais l'histoire des *routes* suivies par l'esprit du " génie inventeur " qui analyse les " petits détails " de la nature. Pour expliquer le monde il ne suffit pas d'en expliquer les faits, encore faut-il les comprendre, c'est-à-dire poser sur eux un discours juste qui ne nécessite aucun appel extérieur à la nature, ce qui serait revenir à un discours non analysable par l'esprit de homme. La compréhension est accessible par le chemin de l'analyse. Il revient très certainement au XVIIIe siècle d'avoir donné à l'histoire des sciences sa dignité philosophique[17].

---

17. M. Foucault, " La vie : l'expérience et la science ", *Revue de Métaphysique et de Morale*, n° 1 (1985), 5.

DES OBSERVATIONS ASTRONOMIQUES VIEILLES DE DEUX SIÈCLES TOUJOURS D'ACTUALITÉ - UN EXEMPLE RELATIF À NEPTUNE

Suzanne DÉBARBAT - Simone DUMONT

1. CATALOGUE D'ÉTOILES DES LALANDE, ONCLE ET NEVEU[1]

Le grand catalogue d'étoiles entrepris à l'Observatoire de l'Ecole militaire vers 1777, à la demande de J. Lalande et avec sa participation, par Lepaute d'Agelet a été continué et achevé, après une interruption de deux années, par M. Le François Lalande, neveu de Jérôme. Les 50 000 étoiles observées, du pôle jusqu'au tropique du Capricorne, jusqu'à la magnitude 10, ont été publiées dans *Histoire Céleste* en 1801 par J. Lalande. Environ 12 000 de ces étoiles ont été calculées et réduites par Mme Le François Lalande, mais n'ont peut-être pas toutes été publiées.

L'observatoire de l'Ecole militaire et son quart-de-cercle

Un premier observatoire a été obtenu par Jeaurat, en 1768, du ministre de la guerre de Louis XV, le duc de Choiseul. Un grand quart-de-cercle mural muni d'une lunette achromatique de sept pieds et demi, qui a été commandé à Bird en 1775 par Bergeret, receveur général sollicité par J. Lalande, est installé en 1778 à l'observatoire. Les nouvelles constructions de l'Ecole militaire ont entraîné en 1786 la démolition de l'observatoire, mais dès 1787, il est reconstruit sur ordre du maréchal de Ségur. Jérôme Lalande qui a acheté ce grand mural le 20 novembre 1786, le destine au nouvel observatoire. En 1789, il fait construire une lunette méridienne achromatique de quatre pieds, pour vérifier les ascensions droites des étoiles. A cette époque, une nouvelle destination est envisagée pour l'Ecole militaire, mais l'astronome Bailly (1736-1793) a obte-

---

1. Sauf indication contraire, toutes les citations sont extraites de *Histoire de l'astronomie depuis 1781 jusqu'à 1802* de J. Lalande.

nu la conservation de l'observatoire ; l'Assemblée nationale, la municipalité et le ministre ont tous concouru à son maintien.

Au mois de septembre 1798 (fin de l'an VI de la République), Lalande a fait installer une nouvelle lunette méridienne, travaillée par Lenoir, avec un objectif de Caroché, dont l'ouverture est de 32 lignes, comme celle du quart-de-cercle. Celui-ci n'est pas dans un plan assez uniforme pour que les passages observés le soient bien dans le plan méridien. La lunette est placée contre un mur épais aux fondations profondes ; utilisée comme instrument des passages, elle permet de déterminer les erreurs du plan du quart-de-cercle ou bien les ascensions droites des étoiles fondamentales des zones observées.

Pour la détermination du temps, les Lalande disposent d'une pendule fabriquée par les frères Lepaute (1709-1789) (1727-1802). Cette pendule a quelquefois gardé la seconde pendant un mois ; si bien que J. Lalande note : " Elle retardait en été de deux dixièmes de seconde par jour ; mais pendant trois mois il n'y avait pas eu sur le mouvement journalier une différence d'un vingtième de seconde ".

Les observations et le catalogue de 50 000 étoiles

Joseph Lepaute d'Agelet, d'abord élève de Lalande avec qui il travaille au Collège Mazarin, puis professeur de mathématique à l'Ecole militaire en 1777, y observe au grand mural dès 1778 et présente à l'Académie ses observations de planètes et d'étoiles en 1780. Jérôme Lalande lui a conseillé d'entreprendre un catalogue d'étoiles. Ses observations sont publiées dans les volumes des *Mémoires de l'Académie* de 1784 à 1789, et plus tard par Lalande dans son *Histoire Céleste*. Il laissera cette entreprise inachevée, étant parti le 1er août 1785 avec l'expédition La Pérouse dont on connaît le sort tragique.

Après la reconstruction de l'Observatoire de l'Ecole militaire, J. Lalande souhaite poursuivre le travail commencé par d'Agelet. La lunette du grand mural lui permet de voir des étoiles jusqu'à la dixième grandeur ; il peut ainsi compléter les catalogues antérieurs, tel celui de 2 884 étoiles que Flamsteed (1646-1719) a fait un siècle auparavant. Depuis cette époque, plusieurs de ces étoiles se sont déplacées ; il sera donc possible de déterminer leur mouvement propre. Certaines avaient peut-être des positions erronées, d'autres ont pu disparaître. Le nouveau catalogue est, selon J. Lalande, un outil indispensable en particulier pour l'étude des étoiles variables et pour la recherche des comètes. Ne pouvant pas entreprendre seul ce travail, c'est son neveu, Michel Le François Lalande, secondé par sa jeune épouse, qui a été le principal observateur. En 1789, il a été aidé par Ungeschick (1760-1790), missionnaire lazariste, et plus tard, à partir du 15 décembre 1797, il est secondé par Burckhardt (1773-1825), jeune astronome allemand.

J. Lalande organise le travail en partageant le ciel en zones de 2° depuis le pôle jusqu'au tropique du Capricorne. Il commence par les étoiles circumpolaires en plaçant le quart-de-cercle mural du côté du nord. Ce qui permet en 1790 de bien placer, dans ce champ d'étoiles, la route de la comète découverte par Caroline Herschel (1750-1848) le 17 avril dans la constellation d'Andromède et qui est passée près du pôle. J. Lalande détaille les résultats : " J'eus dans les six premiers mois, 3 000 étoiles boréales, jusqu'à la 10e grandeur (…). En continuant ainsi, nous devions en avoir 10 000, là où Flamsteed n'en avait pas 600 ; et nous étions sûrs d'avoir, dans toutes les parties du ciel, des points fixes pour les comètes que l'on pourrait découvrir, et de fournir à ceux qui nous suivront, des termes de comparaison pour les étoiles qui changent de place, et pour celles qui disparaissent ou qui paraissent de nouveau… ". La moisson est abondante : " deux ou trois cents étoiles sont le fruit d'une nuit souvent très froide et bien pénible ".

J. Lalande peut annoncer que fin 1794, ils ont plus de 25 000 étoiles, ajoutant que " Mme Le François secondait avec zèle son mari dans ses observations et ses calculs ". En 1797, il annonce 43 400 étoiles et près de 50 000 en 1799. Le travail est achevé en 1800. Alors, M. Le François et Burckhardt revoient les constellations zodiacales, à la recherche de planètes. En 1796, le ministre Bénézech a ordonné l'impression de l'*Histoire Céleste,* publiée en 1801 et qui contient les observations faites à l'Observatoire de l'Ecole militaire et aussi celles de Darquier (1718-1802) à Toulouse de 1791 à 1798.

La réduction des observations

Dans la préface de l'*Histoire Céleste,* J. Lalande écrit : " Le C. Cassini, en publiant les observations de 1785 à 1791, a donné les réductions et les calculs des principales observations ; mais j'ai cru qu'il valait mieux laisser le soin des calculs à ceux qui voudraient faire usage des observations, et qui auraient dans la suite plus de secours, plus de données et des élémens plus exacts. L'astronome qui a besoin d'une observation importante, la calcule et la discute toujours… ". Il estime donc qu'il est important d'observer un grand nombre d'étoiles, pour la recherche des comètes et la détermination de leur route, comme on l'a vu plus haut ; il espère aussi sans doute qu'on trouvera quelque planète dans ce catalogue de même qu'il a trouvé qu'Uranus, découvert par Herschel en 1781, était la 34e étoile du Taureau dans le catalogue britannique, observée en 1690. Ses détracteurs disent qu'il vaudrait mieux avoir moins d'étoiles et y mettre une plus grande précision. Il répond : " ils se trompent ; c'est le grand nombre d'étoiles qui remplit l'objet nécessaire de ce travail : une plus grande exactitude est inutile quant à présent, et le sera longtemps encore ". Cependant le travail de réduction des observations n'est pas vraiment négligé.

J. Lalande sait utiliser les talents : " Il y a des personnes qui ont un goût inné pour le calcul, pour qui calculer est un besoin ; j'en ai rencontré plusieurs, et j'ai tâché d'en tirer parti pour le bien de l'astronomie… ". Dans la correspondance (ms. BOP) qu'il adresse à Flaugergues (1755-1830), astronome et juge de paix à Viviers, il est possible d'en trouver quelques traces. Par exemple :

" 25 nov. 1801
J'ai reçu avec grand plaisir mon cher confrere vos 14 etoiles, et j'attens les autres avec grande impatience pour terminer mon catalogue...
4 decembre 1801
(…) Puisque vous avez peu de tems et que je suis tres impatient je vous prie de m'envoyer les observations d'etoiles que je vous envoyais le 16 oct. M. Chabrol qui vient d'arriver m'offre de les reduire et il le fera dans peu de jours. ne gardez que celles que vous pouvéz finir la semaine prochaine
Salut consideration attachement
Lalande
A Paris le 23 Vendemiaire an 11 (15 octobre 1802)
Puisque vous avez envie mon cher confrère de cooperer encore à mon travail je vous envoie les dernieres étoiles que Mr Vidal a observé à Mirepoix pour les réduire à 1790... "

Flaugergues, passionné d'astronomie et membre correspondant de l'Institut, n'est pas uniquement un calculateur ; mais, grâce à ces calculs, Lalande peut lui obtenir quelques gratifications.

Les réductions des étoiles observées à l'Ecole militaire sont principalement effectuées par Mme Le François. En 1795, " elle avait déjà réduit plus de 1 500 étoiles, et chaque réduction exige trente-six opérations de calcul ; elle a été ensuite à 12 000, et elle continue ". Parmi ces 12 000 étoiles, les principales sont dans le grand Atlas de Bode à qui J. Lalande les a communiquées. Par ailleurs, il en publie une partie dans les *Additions* à la *Connaissance des Temps*. Ainsi, dans la *Connaissance des Temps pour 1797*, il se trouve 1 000 étoiles circumpolaires observées plusieurs fois et réduites à 1790 ; 3 000 autres étoiles sont dans la *Connaissance des Temps pour 1799*. Le catalogue des 50 000 étoiles, que J. Lalande a placé dans son *Histoire Céleste*, a été complété et révisé par F. Baily (1774-1844) ; il a été publié en 1847.

2. Observations de Neptune par " Lalande "

De tout temps, la découverte d'un objet ou d'un phénomène nouveau entraîne des recherches dans les archives passées. Les preuves abondent, y compris l'identification possible de Neptune dans les notes de Galilée (Kowal et Drake, 1980). Les observations des Lalande de la fin du XVIIIe siècle contenaient une nouvelle preuve de l'intérêt de la conservation des archives astronomiques.

Découverte de Neptune

La découverte de Neptune, le 23 septembre 1846, par Galle et d'Arrest à l'Observatoire de Berlin alors dirigé par Encke (1791-1865), à moins de 1° de la position prédite par Le Verrier, a fait, depuis cette époque, l'objet de nombreux articles. Articles de recherche, articles de synthèse, articles scientifiques, historiques, de vulgarisation... Un colloque organisé par le Département d'Astronomie fondamentale/URA 1125 du CNRS, a donné l'occasion d'un nouvel examen de quelques-uns, relatifs à des observations de M. Lalande datant de 1795 (Débarbat, 1996). En effet, comme dans le cas d'Uranus et compte tenu de la magnitude, dès la découverte du nouvel objet, les astronomes ont recherché s'il n'avait pas fait l'objet d'observations antérieures à 1846.

Dans le cas d'Uranus, 19 observations au total avaient été trouvées par différents astronomes, en dernier lieu par Alexis Bouvard (1767-1843), couvrant la période 1690-1771. Ces observations venaient de jouer un rôle capital dans la découverte de la planète qui faisait suite à Uranus dans l'ordre des distances au Soleil : Bouvard, dans ses tables d'Uranus parues en 1821, n'avait pu prendre en compte les observations du passé, celles-ci se révélant incompatibles avec celles menées depuis 1781, pour établir les lois du mouvement.

La même quête d'observations avait été menée à l'issue de la découverte de Neptune, il y a maintenant un peu plus de cent cinquante ans. Sous plusieurs plumes, l'une d'entre nous avait pu lire que Le Verrier avait lui-même identifié deux observations de Neptune, vue comme une étoile et, forte de ces affirmations, l'avait répété. En fait, il n'en est rien et il suffit, pour s'en convaincre, d'examiner les *Comptes rendus de l'Académie des sciences* de l'année 1847.

A la date du 29 mars 1847, une *Note présentée par M. Le Verrier* précise : " Deux observatoires, l'un en Europe, l'autre en Amérique, celui d'Altona et celui de Washington, sont arrivés, chacun de leur côté à un même résultat, dont la constatation physique ne laisse rien à désirer ".

Le Verrier lit alors à l'Académie une lettre de Schumacher (1780-1850), directeur de l'Observatoire d'Altona à propos d'une recherche de Petersen : " Il a trouvé le 17 mars une étoile de l'*Histoire céleste*, qui a disparu du ciel. C'est l'étoile de 7-8 grandeur, observée par Lalande, le 10 mai 1795 ". L'*Histoire Céleste* est celle de Lalande Joseph-Jérôme et Lalande est Michel, neveu du premier qui a mené à l'Observatoire de l'Ecole militaire la majeure partie des observations du catalogue de 50 000 étoiles publié par son oncle en l'An IX de la République. Schumacher ajoutait, dans sa lettre à Le Verrier : " Il est possible que cette étoile fût votre planète ; et je serais bien heureux si mon observatoire vous en avait procuré une observation qui date de 1795 ".

Le Verrier mentionne également qu'il a reçu le *Boston Courier,* numéro du 15 février, arrivé à Liverpool le 16 mars, dans lequel se trouve une lettre de Maury (1806-1873), directeur de l'US Naval Observatory. Le Verrier cite l'article : " M. Walker […] reconnut que la planète avait dû, le 10 mai 1795, se trouver dans le même lieu du ciel qu'une étoile de 7-8 grandeur observée par Lalande. […] La nuit du 4 février fut claire […] l'étoile elle-même manquait ". Le Verrier tire la conclusion : " Il resterait à examiner avec soin si l'astre observé par Lalande en 1795 est bien la planète de 1846 ; […] comme l'ont pensé MM. Petersen et Walker ".

Les *Comptes rendus* du 12 avril 1847 mentionnent une nouvelle lettre de Schumacher : M. Petersen a trouvé encore deux étoiles de Lalande qui ne sont plus au ciel. Mais il faut admettre, chaque fois, une erreur d'une minute sur les temps d'observation ; c'est pourquoi Mauvais (1809-1854) indique, ce même jour : " Pour lever les doutes qui pouvaient rester sur l'existence de l'étoile de l'Histoire céleste indiquée par MM. Petersen et Walker comme pouvant s'identifier avec la nouvelle planète, M. Arago m'avait engagé à vérifier cette position sur les manuscrits qui lui ont été offerts par les héritiers de Lalande, et dont il a fait généreusement don à la bibliothèque de l'Observatoire ; je me suis empressé de m'occuper de cette vérification, et déjà je crois être arrivé à un résultat décisif en comparant la position observée le 10 mai 1795 avec une autre observation faite deux jours auparavant, mais qui n'a point été imprimée, parce que Lalande, en la soumettant au calcul, crut reconnaître qu'elle était entachée d'une double erreur, sur l'instant du passage et sur la hauteur qui, suivant lui, aurait été intervertie avec celle d'une autre étoile ".

Arago (1786-1853) avait, en effet, lui-même après la lecture à l'Académie de la lettre concernant les deux nouvelles observations possibles, fait une déclaration ainsi rapportée : " Au sujet des erreurs de minutes, signalées comme possibles par M. Petersen, M. Arago déclare qu'il a déjà prié un de ses collaborateurs et confrères (M. Mauvais) de se livrer sur ce point à toutes les vérifications désirables, et de consulter pour cela les registres manuscrits de Lalande. M. Arago, à qui ces registres avaient été offerts en cadeau par M. Le chef d'escadron Lefrançais de Lalande, annonce qu'ils sont actuellement déposés à la Bibliothèque de l'Observatoire ".

Neptune dans les carnets dits de Lalande

A la séance suivante, du 19 avril 1847, Mauvais présente une note à l'Académie, *Sur une observation inédite de la nouvelle planète* rappelant les faits : " Dans la séance de lundi dernier […] observation inédite faite à l'Observatoire de l'Ecole militaire, le 8 mai 1795, qui me paraissait devoir servir à lever toutes les incertitudes relatives à l'étoile indiquée par MM. Petersen et Walker, et qui, en même temps, fournirait probablement une observation précieuse de la nouvelle planète, observation qui serait restée à jamais ignorée sans la possession des manuscrits originaux qui nous l'ont conservée […] ". Mauvais poursuit : " Je vais d'abord copier ici les observations originales, afin que les astronomes puissent répéter mes calculs et vérifier les conséquences que j'en tire :

Le 8 mai 1795	Passage observé	Distance zénithale observée
Etoile de 7-8ᵉ grandeur : fil du milieu	14h11m 24ˢ	59°54'40"
Etoile de 7ᵉ grandeur : fil du milieu	14h11m 36ˢ.5	60° 8'17"

Après la première de ces deux observations on lit la petite Note suivante, écrite après coup en marge du manuscrit : " Voyez le 10 mai ; il y a transposition de hauteur et erreur sur le passage de [l'autre] l'étoile suivante ". La seconde observation est celle que j'identifie avec la planète. Pour faciliter les comparaisons, je reproduis ici les deux observations correspondantes, publiées à la page 158 de l'*Histoire céleste* :

Le 10 mai 1795	Passage observé	Distance zénithale observée
Etoile de 7-8ᵉ grandeur : fil du milieu	14h11m 23ˢ.5	60° 7'19"
Etoile de 8-9ᵉ grandeur : troisième fil	14h11m 50ˢ.5	59°54'40"

Mauvais ajoute : "Après la première de ces deux observations on lit en marge : Voyez le 8 mai ; il y a transposition de hauteur avec l'étoile qui est à 59° 54' 40". De plus, on a raturé le troisième fil de cette première observation, il portait 50.5s ; ce nombre a été évidemment reporté plus bas, comme étant le passage au troisième fil de l'étoile correspondant à 59° 54' 40" de distance au zénith ".

Si l'on se reporte au carnet d'observation (ms. BOP), il est aisé de constater que Mauvais a parfaitement bien recopié les valeurs numériques et les indications, que la rature est un trait barrant la valeur 50.5s. En outre, cette observation est suivie d'une observation d'une étoile de 8-9ᵉ grandeur dont le passage observé, au troisième fil, est 14h 11m 50.5s et la distance au zénith observée 59° 54' 40" puis : " voyez le 8 mai ". Dans l'imprimé de l'*Histoire Céleste* le 11m 50.5s est suivi des deux points (:) traditionnels du volume pour signaler une valeur douteuse. Deux corrections en marge, au crayon, d'une écriture plus récente portent l'une sur l'étoile de la Balance, l'autre affecte la valeur de l'imprimé 11m 50.5s pour remplacer 11m par 12m. Comme cet objet suit l'étoile précédente observée à 4h 12m 31.4s, il est plausible de considérer que l'imprimé est affecté d'une erreur d'une minute d'heure. Dans le carnet d'observation, examiné par Mauvais en 1847, la remarque qu'il a faite sur le " reporté plus bas " s'applique à cette ligne pour laquelle la distance zénithale observée est de 59° 54' 40".

Mauvais poursuit : "Au reste, les observations originales ne portent aucun signe d'incertitude et les deux points indiquant le doute, qui se remarquent à la page 158 de l'*Histoire céleste*, n'existent pas dans le manuscrit. Les annotations marginales porteraient au premier abord, à rejeter toutes ces observations comme défectueuses ; mais, en examinant de près la différence qui existe, soit entre les instants des passages, soit entre les hauteurs de la seconde étoile du 8 mai, comparés avec ceux de la première du 10, on ne tarde pas à reconnaître que ces différences sont, à très peu de choses près, le mouvement rétrograde de la planète dans l'intervalle de deux jours, pour le lieu qu'elle devait occuper à peu près à cette époque ".

L'examen du carnet d'observation et de l'imprimé montre qu'il y a peu d'indications du type (:) et que lorsqu'elles existent elles correspondent généralement au mot " dout " ou " douteuse " sauf dans le cas des anomalies repérées les 8 et 10 mai 1795. On remarque aussi, dans le carnet qu'il y a au moins trois écritures différentes. Mais il ne faut pas oublier qu'il y avait deux Lalande, d'autres observateurs et l'épouse du neveu dont il est connu qu'elle effectuait la plus grande partie des calculs nécessaires aux travaux de son " oncle " et père adoptif, et de son mari.

3. NEPTUNE EN CALCULS

Les carnets des Lalande ayant livré, aux astronomes de l'époque, des informations d'intérêt primordial pour la poursuite des études, il restait à démontrer que les 8 et 10 mai 1795, ils avaient bien — sans s'en rendre compte — observé non une étoile mais une nouvelle planète.

Les calculs de Mauvais, Walker, Peirce et les autres

Après son examen des carnets, Mauvais effectue un " calcul rigoureux ", déterminant les positions pour les deux dates à partir des étoiles de la zone observée. Il en tire les valeurs suivantes :

	Temps moyen de Paris[2]	Ascension droite apparente	Déclinaison apparente
Le 8 mai 1795	11h10m 57s	212° 59'35".0	-11°20'39".1
Le 10 mai 1795	11h 2m 55s	212° 56'36".3	-11°19'38".8[3]
Différences		-2' 58".7	+1' 0".3

Mauvais utilise ensuite l'orbite calculée par Walker ce qui lui donne, respectivement, pour l'ascension droite et pour la déclinaison, à chaque date :

8 mai	213° 1'5"	-11°13'5"
10 mai	212°58'5"	-11°12'4"
Différences	-3'0"	+1'1"

---

2. Un correctif de Mauvais, publié peu après, signale qu'il faut lire respectivement : 11h 5m 11s, 10h 57m 9s.

3. A noter que, pour cette valeur, une erreur typographique a fait imprimer h au lieu de ° au nombre 11.

Mauvais conclut alors : " Ces rapprochements sont tellement précis, qu'il me paraît impossible qu'ils laissent aucun doute dans l'esprit des astronomes ". Puis il examine le cas des deux étoiles nouvellement signalées par Petersen ; mais n'en tire pas de conclusion positive.

Dans son étude datée du 10 avril 1848, Walker reprendra ces deux observations de 1795 auxquelles il en ajoute 1 154 autres émanant de différents observatoires du monde entier jusqu'au 27 janvier 1848. Il établit de nouvelles éphémérides de Neptune et forme des (O-C) c'est-à-dire " Observation-Ephémérides ". Les écarts qu'il indique sont respectivement, pour le 8 mai $\Delta\alpha$ = -0".29 et $\Delta\delta$ = +0".79, pour le 10 mai $\Delta\alpha$ = +1".18 et $\Delta\delta$ = +0".31. Ces écarts sont tout à fait normaux dans la série des 1 156 observations traitées où l'on trouve, au hasard des dates, des écarts, pour l'une ou l'autre des coordonnées, qui peuvent atteindre 5", 6", 7" et même jusqu'à 10". Ces deux observations s'insèrent donc bien dans l'ensemble étudié par Walker dans l'année qui a suivi la découverte de Neptune. Il pouvait, à juste titre, être satisfait de son travail.

Polémiques et controverses

Toutefois une petite polémique est intervenue dès avril 1847. Dans les *Astronomische Nachrichten* du 3 avril, Schumacher avait fait connaître la trouvaille de Petersen dans l'*Histoire Céleste* pour le 10 mai 1795. Puis après une étude rapportée le 15 avril, une lettre de Hind (1823-1895) du 13 avril, mentionnée dans le numéro du 1er mai, signale : *After a rigorous reduction, I find the position of the planet to be…*, et il donne à la date du 10 mai 1795 à Paris temps moyen 10h 57m 1s : ascension droite 212° 56' 33.11" et déclinaison -11° 19' 36.61" ; il a mené ses calculs en utilisant huit étoiles dont il indique qu'elles avaient été bien observées par Piazzi (1746-1826) et par Henderson (1798-1844).

A la même date (1er mai), une lettre de Le Verrier du 22 avril est rapportée : " Les registres manuscrits de Lalande sont déposés à l'Observatoire. Mr Mauvais a été chargé de les examiner, à l'occasion de vos communications. Il en est résulté que Lalande avait observé la planète non seulement le 10 mai, mais encore le 8 mai. Mais il avait observé en même temps une étoile peu éloignée, et le mouvement propre de la planète aidant, il en était résulté, au moment de l'impression, une confusion qui avait fait supprimer deux observations le 8 mai. Je les rétablis ici d'après Mr Mauvais.

	fil milieu	Dist. au Zénit
Mai 8 * (7.8^m)	14h11m24s	59°54'40"
Planète	14 11 36.5	60 8 17
- -		
Mai 10 Planète	14 11 23.5	60 7 19
Même *	3e fil	
que le 8	14 11 50.5	59 54 40

Si Mr Petersen veut bien intercaler deux de ces lignes dans son histoire céleste elle deviendra conforme au manuscrit.

L'observation du 10 Mai, qui est imprimée, est marquée de deux points : douteuse. Ces points n'existent pas dans le manuscrit.

Quant à l'étoile de la page 160, la minute y est effectivement surchargée après coup, en sorte que votre hypothèse qu'il pouvait y avoir erreur d'une minute se trouve très probablement fondée.

Mais l'étoile de la page 347 est imprimée conformément au manuscrit ".

Puis le numéro du 20 mai fait état d'un envoi d'Everett qui indique : *Professor Peirce communicated to the Academy the following notice of the computations of Mr Sears C. Walker, who had detected a missing star in the* Histoire Céleste Française, *observed by Lalande on the 10th of May, 1795, near the path of the planet Neptune, at that date, which may possibly have been this planet.* Un peu plus loin : *Mr Walker found that Lalande had twice included the Neptunian region in his sweeps, viz. May 8th and 10th 1795.* (Walker examine donc dans l'*Histoire Céleste* les éléments pour ces deux dates. Il trouve une observation le 10 mai). *Accordingly, he computed the locus of Neptune on the latter night...* Il ne pouvait pas trouver l'observation du 8 mai puisque l'*Histoire Céleste* ne la donnait pas.

L'étude qui suit s'emploie à démontrer que la découverte de Galle et d'Arrest, sur les indications de Le Verrier, n'est qu'un *happy accident*. C'est sans doute cette remarque qui conduit Mauvais à s'adresser à Schumacher, le 5 juillet, dans une lettre publiée au numéro du 12 août : " Vous avez sans doute pris connaissance des résultats des recherches auxquelles je me suis livré pour trouver dans les manuscrits de Lalande, donnés à l'observatoire par Mr Arago, les observations d'étoiles qui peuvent s'identifier avec la planète Neptune. Je crois avoir fait à cet égard un peu plus qu'une simple correction d'épreuve, une simple comparaison d'un imprimé avec son manuscrit. Je crois avoir donné à l'appui de l'identité supposée, les plus fortes probabilités dont on puisse l'entourer, et je pense que les astronomes me sauront gré de mon travail ".

Mauvais donne alors les valeurs qu'il a présentées à l'Académie des sciences, publiées ici avec une erreur, les faisant suivre de la référence à la publication dans les *Comptes rendus* (séance du 19 avril 1847 et *errata* de la fin du cahier du 10 mai). Mauvais termine par " Je vous serais reconnaissant, Monsieur, si vous vouliez bien donner une petite place à ces résultats, dans votre recueil astronomique ".

D'un autre côté, en France même, Babinet (1794-1872) attaque Le Verrier dès 1848 précisant " L'identité de la planète Neptune avec la planète théorique (…) n'est plus admise par personne… ". Alors que Le Verrier reçoit le soutien d'Adams (1819-1892) lui-même, les compliments des collègues britanniques, de Struve (le fils, 1819-1905), Liais (1826-1900) — encore en 1866 — rédige un pamphlet contre " Le travail du calculateur français (qui) fut pour lui

(Galle) ce qu'on raconte qu'avait été pour Newton la pomme tombant d'un arbre… ". Pourtant, de l'autre côté du *Channel* et depuis 1847, hommage avait été rendu au " calculateur français " : une mèche de cheveux de Newton lui avait été offerte, via Mme Le Verrier, accompagnée d'une lettre le considérant comme " le plus illustre de ses disciples ".

En fait cette polémique n'était due qu'à des inimitiés personnelles. Le fait que Adams ait obtenu indépendamment un résultat similaire à celui de Le Verrier assurait d'office aux deux astronomes la validité de leurs analyses.

Nombre d'étrangers et non des moindres, venus notamment de Grande-Bretagne, ont participé — sans aucune réticence — aux célébrations qui ont entouré le centenaire de la découverte de Neptune en octobre 1946 à Paris. La revue *l'Astronomie* en a rendu compte dans sa livraison de novembre-décembre de la même année incluant les discours des différentes personnalités et la conférence, fondamentale, prononcée par Danjon à la Sorbonne, dans laquelle la plupart des auteurs ultérieurs ont puisé sans le citer dans leurs références.

Les orateurs ont développé différents aspects de la découverte, y compris les observations menées par Challis (1803-1882) à Cambridge qui a observé, sur les indications de Adams mais sans s'en apercevoir, la planète Neptune les 4 et 12 août 1846. Ces observations feront partie de celles recherchées, aux côtés de celles de M. Lalande ; il y aura aussi celle que Hind a fait connaître à Le Verrier et que ce dernier rapporte à la séance de l'Académie des sciences de Paris du 25 mars 1850 : " Je vous adresse deux positions de Neptune, résultant d'observations comprises dans les zones de Lomont (Munich). Elles n'ont pas été jusqu'ici, autant que je puisse le savoir, notées comme appartenant à cette planète ".

La première a été prise le 25 Octobre 1845 ; elle se trouve dans la zone 338 à 21h 42m 43.1s ; la planète y est notée comme une étoile de neuvième grandeur. La seconde observation se trouve dans la zone 379, le 7 Septembre 1846, à 21h 54m 24.9s ; et la planète y est notée comme étoile de huitième grandeur. J'ai réduit ces observations, à l'aide des catalogues d'étoiles de Greenwich et d'Edimbourg, et j'ai trouvé les positions apparentes suivantes de Neptune :

	TM de Greenwich	AR	Distance pôle nord
1845. Octobre 25	6h 40m15s	21h42m42.48s	104°14'23.0"
1846. Septembre 7	19 1 57	21 54 44.51	103 16 21.8

La seconde de ces positions a été prise après que vos mémoires de Juin et Août 1846 avaient paru dans les *Comptes rendus* ".

4. NEPTUNE UN SIÈCLE ET DEMI APRÈS SA DÉCOUVERTE

Les sondes Voyager, lancées par la NASA, ont eu l'occasion de faire connaître la surface d'Uranus et celle de Neptune, deux planètes aux couleurs bleutées. Leur nature physique a pu être étudiée. Mais pour pouvoir le faire il a fallu établir les éphémérides devant permettre d'assurer le passage au plus près. Ces éphémérides, établies au Jet Propulsion Laboratory, sont fondées (Standish et Williams, 1990) sur un grand nombre d'observations collectées et rapportées au même système de référence, constituant ainsi une base de données fondamentales pour les études tant théoriques qu'appliquées.

A l'occasion de l'exposition " La mesure du ciel - De la plaque photographique aux techniques spatiales " dont un thème concernait la découverte de Neptune, grâce aux cartes célestes établies à l'époque par Bremiker (1804-1877), les observations de M. Lalande de 1795 sortaient de l'oubli. L'attention de Standish ayant été attirée par l'ancienneté de ces données, il en entreprenait alors l'étude présentée à Cambridge en 1993 (Standish, 1994) ; trois pondérations étaient expérimentées conduisant, en ascension droite, aux écarts suivants (cas 1 et 3) :

	Da (o-c)1	Da (o-c)3
1795 Mai 8	-0.61s	-0.24s
Mai 10	-0.56s	-0.29s

Le système 3 correspond à un poids élevé donné aux observations de Lalande, tandis que pour 1 elles ne sont pas prises en compte. Cette dernière valeur est à comparer à celle obtenue (Chollet, 1996) en formant des o-c (respectivement -0.52s et -0.38s) où o est la valeur de Mauvais du printemps 1847 et c la valeur calculée à partir de la théorie VSOP82 (Bretagon, 1982). A rapprocher aussi les o-c de Walker donnés dans son étude de 1849 : pour le 8 mai 1795, -0.29" soit -0.02s ; pour le 10 mai, +1.18" soit +0.08s. Standish remarque, dans son étude, que la prise en compte des observations de Lalande en 3 montre : *the obvious signature forced into the other observations by the distorsion of the ephemerides in order to even partially fit Lalande's observations.*

Dans la même perspective d'une comparaison des observations et des éphémérides, il convient de mentionner les observations de la décennie écoulée menées aux deux meilleurs cercles méridiens de notre époque, celui de l'Observatoire de Bordeaux et celui (anglo-danois) des Canaries. La comparaison avec les éphémérides DE200 et DE403 du Jet Propulsion Laboratory montre que DE403, plus récente, améliore notablement la cohérence pour les ascensions droites ; mais qu'en déclinaison le changement n'est pas sensible. A noter que les écarts, par rapport à l'une ou à l'autre de ces éphémérides, atteignent 1". A noter aussi que les valeurs trouvées par Chollet, par rapport à VSOP82, et par Walker, par rapport à ses éphémérides, pour les observations de M. Lalande, sont respectivement pour cette coordonnée :

	Dd (o-c)C	Dd (o-c)W
1795 Mai 8	+0.9"	+0.79"
Mai 10	-0.0"	+0.31"

valeurs qui s'insèrent parfaitement bien dans les limites de la dispersion des observations menées entre 1985 et 1995.

5. QUELQUES CITATIONS EN FORME DE CONCLUSION

" Les nouvelles planètes qui existent peut-être, sont un objet également important de notre nouveau travail. M. Herschel en a découvert une par hasard, en 1781 ; et lorsqu'on en découvrira quelque autre, on la trouvera dans nos cinquante mille étoiles, et l'on aura tout de suite de quoi établir la durée de sa révolution ". (J. Lalande, préface de son *Histoire Céleste*, 1801).

" Je laisse au temps à venir le soin de faire connaître, si la difficulté de concilier les deux systèmes tient réellement à l'inexactitude des observations anciennes, ou si elle dépend de quelque action étrangère et inaperçue, qui aurait agit sur la planète ". (Alexis Bouvard, *Tables du mouvement... d'Uranus*, 1821).

" Cela tient-il à une perturbation inconnue apportée dans les mouvements de cet astre par un corps situé au-delà ? Je ne sais, mais c'est du moins l'idée de mon oncle ". (Eugène Bouvard, Lettre à Airy, 6 octobre 1837).

If it be the effect of an unseen body, it will be nearly impossible ever to find out its place. (Airy, réponse à la lettre d'Eugène Bouvard, 12 octobre 1837).

" Je pense qu'un moment viendra où la solution du mystère Uranus sera peut être bien fournie par une nouvelle planète, dont les éléments seraient reconnus par son action sur Uranus et vérifiés par celle qu'elle exerce sur Saturne ". (Bessel, lettre à Humboldt, vers 1845).

" Jamais découverte scientifique n'a fait plus de sensation, et à plus juste titre ". (Otto Struve, octobre 1846).

" M. Le Verrier a aperçu le nouvel astre sans avoir besoin de jeter un seul regard vers le ciel ; il l'a vu au bout de sa plume. (…). Ainsi la découverte de M. Le Verrier, est une des plus brillantes manifestations de l'exactitude des systèmes astronomiques modernes. Elle encouragera les géomètres d'élite à chercher avec une nouvelle ardeur, les vérités éternelles qui restent cachées, suivant une expression de Pline, dans la majesté des théories ". (Arago, *Mémoires de l'Académie des sciences*).

" Ce succès doit nous laisser espérer, qu'après trente ou quarante années d'observations de la nouvelle planète, on pourra l'employer, à son tour, à la découverte de celle qui la suit, dans l'ordre des distances au Soleil ". (Le Verrier, *Connaissance des temps pour 1849*, publiée à la fin de 1846).

BIBLIOGRAPHIE

Manuscrits de la Bibliothèque de l'Observatoire de Paris (ms. BOP) :

1058 II, lettres de J. Lalande à Flaugergues.

C5bis, n° 23, carnet d'observation de M. Le François Lalande.

Articles cités :

P. Bretagnon, " Théorie du mouvement de l'ensemble des planètes. Solution VSOP82 ", *Astronomy and Astrophysics,* 114 (1982), 278-288.

F. Chollet, communication personnelle, 1996.

A. Danjon, " La découverte de Neptune ", *l'Astronomie* (nov.-déc. 1946), 255-278.

S. Débarbat, " A mi-chemin des *Mécaniques Célestes* de Laplace et de Tisserand, une brillante retombée des mathématiques appliquées ", *Journées 1996 Systèmes de référence spatio-temporels,* Observatoire de Paris, Ed. Capitaine, Paris, 1996.

Ch. Kowal et S. Drake, " Galileo's Observations of Neptune ", *Nature,* 287 (1980), 311-313.

E.M. Standish, " Meridian circle observations of the planets ", in Morison and Gilmore (eds), *Galactic and solar system optical astrometry* (Cambridge, Cambridge University Press, 1994), 253-262.

E.M. Standish et J.G. Williams, " Dynamical Reference Frame in the Planetary and Earth-Moon Systems ", in J. Lieske and V.K. Abalakin (eds), *Inertial Coordinate System on the Sky* (Dordrecht, Kluwer Pub., 1990), 173-181.

FIGURES

Fig. 1 - Lalande, buste attribué à Houdon (1741-1828). *Observatoire de Paris.*

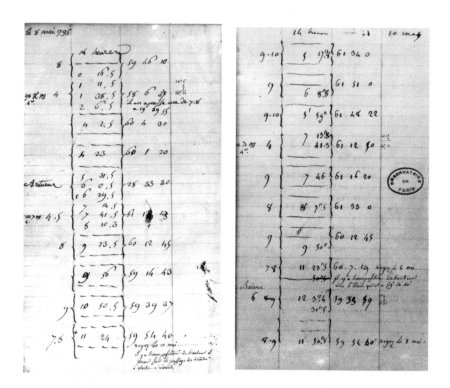

Fig. 2 - Deux pages extraites d'un carnet d'observation de l'Observatoire de l'Ecole militaire ; pour chaque étoile, la première colonne donne la magnitude (7.8 signifie entre 7 et 8). *Bibliothèque Observatoire de Paris.*

Fig. 3 - Reproduction d'un projet de plafond de Dupain (1847-1933) pour la Rotonde Est de l'Observatoire de Paris. Le Verrier y est représenté, sur la terrasse supérieure du Bâtiment Perrault, montrant la direction de Neptune. *Observatoire de Paris.*

Sir Harold Jeffreys, Pioneer of Modern Planetology

Alexander V. Kozenko

To talk about Jeffreys is both easy and difficult. Easy because he was a remarkable scholar belonging to the famous School of English theoretical physics and a great worker. In his long life (his last paper was published at 95 !) he did so much, particularly in geophysics, astronomy and applied mathematics, that even a simple enumeration of his works would take up too much time in this report. Difficult, because the name of Jeffreys is not linked with a sensational hypothesis, such as, for example, the hypothesis of continental drift which is associated with the name of the German meteorologist and geophysicist A. Wegener.

Jeffreys was able to produce a contribution to geophysics comparable to if not greater than that of a large Institute. The work of Jeffreys in the middle of the 20th century determined the face of geophysics. But many of the papers by Jeffreys were on the borderline between geophysics and astronomy. In his paper on *The Relations between Astronomy and Geophysics* he wrote : " I am often asked (quite kindly) how it happens that a geophysicist holds the Plumian Chair of Astronomy in succession to Sir Arthur Eddington. There was, in fact, a great precedent. Eddington's predecessor, Sir George Darwin, was one of the greatest geophysicists, and it was largely his work that first attracted me to geophysics. Of the four main volumes of his collected works, the first is entirely concerned with tidal theory ; the others are about equally divided between astronomy and geophysics. The relation between the two subjects has long been recognized in Great Britain by the Royal Astronomical Society. The Earth is a planet, and it is the one that we know most about ".

On the boundary between astronomy and geophysics there is the subject of gravimetry. Jeffreys did much work in this field, in particular on determination of the parameters of the gravitational field, figure of the Earth and planets, the nature of isostasy, and considering the interpretation of the results of the variation of the force of gravity in relation to questions of gravitational research.

Jeffreys actively studied the theory of tides. In 1928 he discovered the influence of zonal tidal deformation on variation of the greater moment of inertia of the Earth. This required a variation in the velocity of rotation. His theoretical results were later confirmed by observations.

Already in 1926 Jeffreys first considered Earth tides in a two-layered model of the Earth (a uniform core and mantle in which the rigidity of core was taken as zero).

However this approximate model agreement with data from the motion of the Pole was not satisfactory, further improvements of the theory required the use of complicated rheological models for stress-strain relation (the simplest of these would be Hooke's law). Looking for adequate rheological description of the Earth's interior Jeffreys advocated the modified law of Lomnitz.

Jeffreys made a substantial contribution to the theory of the motion of the Moon, the dynamical problem of the Earth Moon system. Owing to the inelasticity of the Earth there is a relation between the slowing of the Earth's rotation and the secular acceleration of the Moon. Nearly all the retardation comes from tidal friction and in particular in shallow seas. From Jeffreys' estimate two thirds of that required could come from the Behring Sea. Further calculations showed that the total effect of all the oceans was of the same order as that round by Jeffreys.

The present time is the epoch of the unification of the physics of the Earth with planetology and cosmogony. In fact geophysics gives the answer to many problems of the origin and evolution of bodies of the Solar System. It is noteworthy that Jeffreys was ahead of his time in thinking and working in such a framework.

Noticing in his work the importance of studying other planets as well as the Earth and taking knowledge of the Earth as fundamental to knowledge of the others, Jeffreys stand at the source of planetology. In 1937 in *The density distributions in the Inner Planets* he attempted to use the data available to test geophysical hypotheses about the variation of materials in the interior of the Earth. Moreover on the basis of tables for the density of the Earth he constructed the first model of the interior of Venus. From this he concluded that in Venus, as in the Earth, the mantle and core differ in composition. In this work he also gave models of Mars and Mercury. Jeffreys wrote : " So starting from the Earth, we go to the Moon, back to the Earth, and again to the Moon and on to Mercury... "

Jeffreys also did pioneering work in constructing models of the giant planets. Before his 1923 paper on *The Constitution of the four great planets* it was supposed that they consisted of hot gas. He tried to determine the surface temperature and came to the conclusion that probably for these bodies the ordinary equations of state of a liquid and rigid body applied. " If the outer planets were as hot as 300 K they would have had time to radiate away any conceivable original store of heat. It is concluded that the present temperatures must be maintained by solar radiation and therefore very low. It is suggested that Jupiter, Uranus and Neptune may be mostly ice, but the density of Saturn is too low for this " wrote Jeffreys.

In 1934 R. Wildt, on the basis of Jeffreys' results, came to the conclusion that the outer layers of these planets were water. This was confirmed in 1951 in the works of V.G. Fesenkov and A.G. Masevich in Russia, Ramsey in England and De-Marcus in the USA. Theories of the inner layers of giant planets of recent date are in the work of V.N. Zharkov and V.P. Trubitcin in Russia.

The interest of Jeffreys in planetary cosmogony shows that he strikingly realized that it was not possible to construct a realistic evolutionary model of the Earth and planets without solving the problem of their origin. On the other hand, investigations in comparative planetology gave indications to this fundamental problem of natural science. Hence for the whole of his life Jeffreys was interested in the origin of the Solar System. Although towards the end of his life he used to say that he did not believe any of the existing theories.

In his first cosmogonic investigations Jeffreys tried to construct theory of the formation of a planetary system on the lines of then ruling ideas of James Jeans. The famous historian of astronomy Stephen G. Brush wrote " There is a superficial similarity between the American pair, Chamberlin and Moulton, and the British pair, Jeffreys and Jeans. James Hopwood Jeans (1877-1946) like Moulton, was primarily interested in the mathematical aspects of astronomy and physics, but pursued original research for only about 25 years ; Harold Jeffreys (1891-1989), like Chamberlin, is more concerned with applications to the history and present state of the Earth, and has been active in science into his eighties. Jeans and Jeffreys improved the Chamberlin-Moulton theory by putting it on what appeared to be a superior astronomical and physical foundation ; they retained the dualistic feature (subsequently abandoned) while discarding the planetesimal feature (later revived) ".

In the late 40s and early 50s Jeffreys gave a general analytical survey of this problem, considering side by side tidal and catastrophic theories, the hypotheses put forward during the war by C. von Weizsacker in Germany and O.J. Schmidt in Russia playing an important role in modern cosmogony. He also put the question of the relation between the cosmogonic scenario and the structure and original state of the planets. He believed the original state to have been hot, effectively hot, and that is today the opinion of most investigators.

SCHUSTER-SCHWARZSCHILD'S AND MILNE-EDDINGTON'S VIEWPOINTS ON THE PHYSICAL STRUCTURE OF STELLAR ATMOSPHERES

Léo HOUZIAUX

I. INTRODUCTION

In the early 1930s, astrophysicists had various conceptions about the structure of stellar atmospheres.

Basic ideas about the physical nature of the outermost layers of a star arose from the interpretation of Kirchhoff's laws on the appearance of spectral lines i.e. dark lines due to the absorption or to the scattering of a continuous radiation by vapors of various elements. As the stellar spectra (and, in particular, the solar spectrum) usually showed such absorption lines, a thin layer of diluted (generally metallic) vapors had to rise above the stellar " surface " (called the *photosphere*), emitting a black body continuous radiation. This thin gaseous atmosphere had been called the *reversing layer,* as it was responsible for the " reversed " appearance of the spectral lines.

Such a picture of the external layers of the Sun and stars arose from an interpretation of Kirchhoff's laws : while the spectra of nebulae (like the Orion nebula) consisted of bright lines, the spectra of the Sun and stars were full of dark lines which were supposed to originate in a stratum of relatively cool gas overlying the incandescent photosphere. If seen alone, that thin layer of cool gas should reveal, like in the case of nebulae, a spectrum consisting of bright lines at the same location as the absorption lines, but appearing against a dark background.

The phenomenon was actually observed by C.A. Young during a total solar eclipse in 1870 (Young, 1893). He reports in his book, *The Sun*, his observations as follows : " As the Moon advances, making narrower and narrower the remaining sickle of the solar disk, the dark lines of the spectrum of the most part remain unchanged... But the moment the Sun is hidden, through the whole length of the spectrum, in the red, the green, the violet the bright lines flash out by hundred and thousands, almost startingly, like stars from a bursting rocket head, and as evanescent, for the whole thing is over in two or three seconds ".

From such a striking observation, it was inferred that this region, obscured in two or three seconds by the Moon was indeed the *reversing layer* postulated by the appearance of dark lines in the solar spectrum.

The important papers published in America by Sir Arthur Schuster and in Germany by Karl Schwarszchild were based on this conception. Schuster (1851-1934) demonstrated in his famous 1905 paper, " *Radiation through a foggy atmosphere* ", that the flux F_v emerging from a homogeneous scattering plane-parallel atmosphere of height H was related to the incident radiation F_{v_o} through the relationship :

$$F_v = F_{v_0}/(1 + \tau_v), \tag{1}$$

α_v being the atomic absorption coefficient at frequency v, N the particle density and hence

$$\tau_v = \alpha_v \, N \, H \tag{2}$$

If the line would have been caused by absorption of radiation (without remission), the corresponding relationship would have been :

$$F_v = F_{v_0} e^{-\tau_v}$$

Hence the so-called *Schuster-Schwarzschild model* of a stellar atmosphere is made of a discrete photospheric surface responsible for a black body emission of continuous radiation, on top of which sits an homogeneous and isothermal layer of cooler gas.

As in the solar spectrum, the most prominent absorption lines due to neutral Sodium and ionised Calcium were resonance lines, the most common mechanism of interaction between the atoms in the reversing layer and photospheric radiation was thought to be scattering without alteration of the photon's frequency (a phenomenon later called *coherent* scattering). For a reversing layer optically deep enough, the photospheric central flux could then be totally depleted. It was however shown that in the solar spectrum, even the center of the strongest lines showed a non-zero intensity, while for hotter stars, the strongest lines never exceeded depths amounting to 40-50 % of the local continuum flux.

Such observations could be explained by Milne's concept of line formation under the conditions of *local thermodynamic equilibrium*. In this picture, the emission of radiation anywhere in the star's external layers is governed by Kirchhoff's law, which means that the emissivity j_v (energy emitted by second by a gram of stellar material) is linked to the absorption coefficient κ_v (in cm^2 per gram) by the relationship :

$$j_v = \kappa_v \cdot B_v(T) \tag{3}$$

and hence, is characterized by the value of the local temperature T. As the probability that a photon escapes from the atmosphere is $exp(-\tau_\nu)$, where τ_ν is the optical depth of the layers above the considered point in the stellar atmosphere, only photons originating in places where τ_ν is about 1 (as $e^{-1} \approx 0.5$) can escape and then form the emergent spectrum. In such a scheme, the emission of either continuous or line radiation is treated exactly in the same way. The intensity of the emergent spectrum thus depends on the opacity of the stellar atmosphere at each wavelength. When the atmosphere is opaque, as only photons in the outer layers will escape, the emissivity j_ν will correspond to the local temperature T of this layer, while in more transparent regions of the spectrum, the emissivity will correspond to the temperature of a deeper region in the atmosphere. In particular, at the center of very strong lines, the intensity will correspond to the temperature T_0 of the outermost surface of the star. As the lines appear in absorption, this explanation postulates the existence of a temperature gradient in the atmosphere, the temperature decreasing towards the surface of the star. Schwarzschild in 1916 (Schwarzschild, 1916) demonstrated that such a gradient depends on the mode of energy transport in the atmosphere and, in comparing his computations with the limb darkening in the Sun, he showed that the energy transport is by radiation transfer.

So, such a concept could explain why the central depths of even the strongest lines could not exceed some percentage of the local continuum, as it was linked to the ratio of the surface temperature to the temperature existing where the local continuous spectrum was formed. It postulated also that the continuous spectrum and the line spectrum were both formed in a single atmosphere and that a discrete photosphere was an unnecessary concept. Furthermore, as the continuous opacity of an atmosphere varies with wavelength, it explains also that in some spectral regions the strong lines may be very deep and in other regions shallower. Milne's ideas, together with a suggestion by Eddington for solving the equation of radiative transfer and establishing a temperature law $T(\tau)$ gave rise to another model called the Milne-Eddington model atmosphere, which has the following characteristics :

1. The atmosphere is considered as plane-parallel (as in the S.-S. model). This approximation results from the extreme narrowness of the stellar atmosphere versus the stellar radius (in the Sun, about 200 km compared to the solar radius of $6.9 \ 10^5$ km).

2. The temperature law is given by the solution of the transfer equation in a grey atmosphere in radiative equilibrium using the Eddington approximations :

$$T^4(\tau) = T_0^{\ 4}[1 + (3/2)\tau]$$

3. The emissivity j_ν is computed using Kirchhoff's law.

4. The ratio of the absorption coefficient in a line and in the local continuum is independent of the optical depth τ (In practice a mean optical depth τ_R is adopted, computed using the so-called Rosseland mean mass absorption coef-

ficient κ_R). This assumption was introduced to facilitate the solution of the transfer problem in the lines.

The main problem to be solved in the field of stellar atmospheres was to derive from spectral data the chemical composition of the atmosphere. Prior to the work of D. Menzel (1901-1976) (Menzel, 1924) and Cecilia Payne (1900-1979) (Payne, 1925), little was known on the subject and it was thought that, for instance, Iron was a major component of solar-type stars. In spite of rather drastic simplifications on the physics of ions, Payne succeeded to obtain the essential result that the various aspects of the stellar spectra was essentially an effect of variation of atmospheric temperature and not of chemical composition, as, for most of the elements, the relative abundances were close to their Earth crust value, except for hydrogen and helium for which she suspected that values were of the order of one million times the metallic vapor abundances.

This very important result was confirmed by the work of H.N. Russell, Walter Adams and Charlotte Moore during the years 1925-1928. Using the solar line intensities collected by Rowland (1848-1901), and which scaled from -3 to 40 for strong lines, Russell showed that they represented variations up to one million for the number of atoms producing such lines. This was a considerable advance on Miss Payne's work, who, like Saha and Fowler, relied upon the marginal appearance of some lines, a concept largerly depending on the quality of the spectroscopic observations. Russell used in fact a representation of the atmosphere based on the Schuster-Schwarszchild model and his relationship between the Rowland intensities and the number N of atoms in the lower level " above the photosphere " is a prefiguration of the method of the curve of growth analysis soon to be developed. In fact, the inventor of the word *curve of growth*, Marcel Minnaert (1893-1970), replaced in Russell's relationship between the empirical Rowland intensities R by a physically much better defined quantity called the total intensity or *equivalent width* of the line, which was then related to the number of atoms per unit volume in the lower state of a multiplet. Hence the work of Minnaert and his collaborator Mulders was much more to transform the Rowland intensities, depending on the wavelength range into physically meaningful data. Minnaert and Mulders showed that the relationship between the equivalent width of the lines and N had indeed the shape predicted by Wilhem Schütz (1930). They also showed that the line intensities were linked to damping phenomena much more important than those predicted by Lorentz's classical work and should then be assigned to the effect of collisions.

II. Use of the Schuster-Schwarzschild model for the analysis of stellar spectra

Coudé spectrograms of the Mount Wilson 100-inch coudé spectrograph provided Russell and Adams with a wealth of information on line intensities in the bright stars of types from A to M and these authors performed a first differential analysis of the chemical composition relative to the solar composition, assigning to each stars parameters describing their atmosphere as a S.-S. model, which led them to not very successful conclusions.

The same model was used by Albrecht Unsöld (1905 -1996) to analyse the profiles and intensities of the Na I and Ca^+ solar lines formed by scattering and he showed them to be due to absorption coefficient of Lorentzien shape. Unsöld deduced a value of one millionth of an atmosphere for the electron pressure in the solar atmosphere : such significant results were confirmed by a later contribution by Russell which became very famous (*Russell's mixture*) on the composition of the solar atmosphere, where he established the overwhelming importance of Hydrogen over the metals, and such things as the peak abundance of lighter elements like oxygen, carbon and nitrogen, the importance of what we call now the iron peak elements and the tendency for atomic species with even numbers to be about ten times more abundant than elements with odd atomic numbers. Hence, astrophysics of that period held in great consideration the atmospheric model that led to such fundamental results.

The success of Russell's analysis of the solar atmosphere led in the U.S. to a considerable confidence in the S.-S. model, although difficulties remained in the analysis of solar-type stars until the ion H⁻ was recognized to be main opacity source in these atmospheres. The theory of the curve of growth in S.-S. model was given by D. Menzel (1936) and used for the analysis of stellar spectra by a number of astrophysicists, including L.H. Aller as late as 1946 (Aller, 1946).

In Europe, this type of model was also widely in use, especially since the success of the method of coarse analysis (*grobanalyse*) for the B0 V star τ Scorpii by Unsöld. Unsöld however, remained a long time in the opinion that the use of a more sophisticated model analysis would not really improve the accuracy of the derived abundances. However, he soon introduced some improvements in the methodology via results obtained in Holland during the 1930s.

III. THE INFLUENCE OF MINNAERT'S AND PANNEKOEK'S WORK ON THE SUN

Instrumental developments at Utrecht after the first world war led to an accurate spectrometric recording of the solar spectrum and induced new theoretical developments both at Utrecht and at Amsterdam to analyse these results. We already mentioned the introduction by Minnaert of the *curve of growth* concept. In fact, already in 1930, Antoon Pannekoek (1873-1960) published two important papers on the " Theoretical contours of absorption lines " (Pannekoek, 1930) based on Eddington and Milne results. These results were used by Minnaert, who published in the Zeitschrift für Astrophysik (Minnaert, 1935) a paper on the analysis of the external wings of the strong Fraunhofer lines. In this paper, he gives several formulas based on Schuster's relationship (1), and shows that Pannekoek's lengthy calculations lead only to minor improvements with the observations.

In this paper, he also gives an empirical relationship that reproduces the wings " up to 55% of residual intensity ", which reads :

$$\frac{i}{i_0} = \frac{[1 + 26(10)^{-2}(\sigma/\kappa_0)]}{[1 + 86(10)^{-2}(\sigma/\kappa_0)]}$$

where σ and κ_0 are the absorption coefficients per gram of stellar material in the line and in the continuum. If we now compute the line depth R :

$$R = 1 - (i/i_0),$$

this relationship becomes :

$$(1/R) = [1/6(10)^{-1}(\sigma/\kappa_0)] + 66(10)^{-2}$$

which has the usual shape of " Minnaert's interpolation formula " widely used thereafter in the astrophysical literature. In the particular case above, it does not represent the profile of the whole line ; it is of course not intended to, since it is supposed to give a proper fit only for the line wings.

In fact, this formula has been popularised by Unsöld, in his treatise *Physik der Sternatmosphären*, where it is intended to represent the whole line profile and where its usual aspects appears for the first time as :

$$1/R = 1/R_c + 1/NH\alpha_v \tag{4}$$

where $NH\alpha_v$ is the column opacity of the reversing layer at frequency v in the line and where R_c has however a different meaning. It represents the largest depth that a line can reach when the opacity at the line center is very large. In the case of the S.-S. model, where lines are formed by scattering, $R_c \longrightarrow 1$. However, when the lines are formed by absorption, R_c remains smaller than 1. In the case of the M.-E. model, R_c is easily computed : in the case of pure scattering, its limit is 1, as in the case of the S.S. model ; if the line is formed by absorption,

$$R_c = (2/3)\beta_0/1 + (2/3)\beta_0,$$

a value rather close to

$$R_c = \frac{1}{1 + \sqrt{3}/\beta_0}$$

obtained with the S.-S. model.

In these relationships, β_0 is a function of the frequency and surface temperature and of the ratio κ_{mean}/κ_v the mean absorption coefficient to the absorption coefficient at the frequency v .

Because of the facts that the line profiles computed with either models gave very similar results (*Physik der Sternatmosphären*, 1st edition, fig. 79), Unsöld was convinced that further refinements were unnecessary and that the interpolation formula (4), was accurate enough for practical purposes. He states : " Generally speaking, one has to admit that the interpolation formula satisfies all the reasonable expectations that one might expect from it ". This is why he developed a general theory of the curve of growth starting from this formula.

In fact, " Minnaert's interpolation formula " can be exactly derived from the solution of a " Milne-Eddington " atmosphere (see Appendix). In this computation, the parameter R_c is found to be :

$$R_c = (2/3)\chi\ \kappa_R\ /[\kappa_\nu + (2/3)\chi\kappa_R]$$ (5)

where

$$\chi = (3/8)(h\nu/kT_0)exp(h\nu/kT_0)/[exp(h\nu/kT_0) - 1]$$ (6)

and κ_R , κ_ν are the Rosseland and monochromatic continuum absorption coefficients (the ratio of which is constant with depth in the M.-E. model).

Then (see the Appendix), the Minnaert formula may be written :

$$(1/R) = (1/R_c) + (\rho\ \kappa_\nu/R_c N\alpha_\nu)$$ (7)

Guido Münch (Münch, 1958) proved in 1958 the " Minnaert formula " would give for high temperature stars profiles systematically in error.

In the meantime, Pannekoek, starting from the principle that " the test of a theory is not complete until, numerically, its consequences have been worked out to the limit imposed by the accuracy of the observations ", undertook to solve the problem with numerical methods appropriate to each situation. Unsöld considered these methods as " laborious ", but, as Mustel noticed, the " laborious " character of these computations is more than compensated by the security of the results. We know now that the current way of handling those problems, is (thanks to the computers) exactly in the line of Pannekoek's work. Let us recall that Pannekoek's work inspired Minnaert's interpolation formula.

On the other hand, in the United States, the Milne-Eddington model was soon recognized to lay on firm physical grounds ; corresponding curves of growths were computed by Wrubel (1949, 1950) and used by a number of authors, including J.L. Greenstein (differential analysis of solar-type stars) and Aller himself until more refined analyses using model atmospheres appeared in the literature (A. Underhill, M. Rudkjöbing,...).

In the various treatises, different attitudes prevailed.

While Unsöld sticks to his S.-S. model and his " Grobanalyse " even in the 1955 revised edition of the *Physik der Sternatmosphären*[1], Aller gives a fair accounts of the various possibilities in his book *The atmospheres of the Sun and Stars* and used only Wrubel's computations for applications of the curve

---

1. He barely devotes three lines in small types to the work of Wrubel, which he qualifies as " umfangreichen Rechnungen "

of growth method. Some authors, like Hack and Struve in their *Stellar Spectroscopy*[2] give tables where the various methods (Menzel, Unsöld and Wrubel are given in parallel. In his *Théorie générale des atmosphères stellaires*[3] Daniel Barbier (1907-1965) assigns both to Unsöld and to Menzel (without mentioning references), the Minnaert formula and uses Wrubel's curves of growth.

IV. SU-SHU HUANG'S REFLECTIONS ON THE CURVE OF GROWTH.

In a study of the atmosphere of Maia, Huang and Struve (1956) recognize that the independent variable used in the curves of growth derived from the S.-S. and M.-E. models are " conceptually different ". But they also recognize that the concept of the reversing layer is sometimes useful and easy to use. But " confusion sometimes arises as to how to take into account the continuous opacity . Rigorously we cannot make such a conversion ". Huang and Struve suggest that an approximate way of connecting the independent variables based on the two completely different concepts. Huang suggests that in the opacity NH α_ν occurring in relation (4) *NH*, the number of absorbing atoms contained in a vertical column to unit cross-section above the photosphere can also be interpreted as the as the number of atoms per unit optical thickness in the neighbouring continuum, in the sense of the M.-E. model, since the continuum is supposed to originate from a layer with $\tau_\nu = 1$. In order to take into account the variation in the continuous opacity, one may write the relation between $NH(\lambda)$ and the continuous opacity $\kappa(\lambda)$ according to the M.-E. model :

$$N(\lambda)/N(\lambda_0) = \kappa(\lambda_0)/\kappa(\lambda)$$

and then interpret the *N*'s in the S.-S. model, that is that the thickness of the reversing layer is inversely proportional to the continuous opacity if everything else remains unchanged. This explanation is advanced by Huang in order to retain the idea of the reversing layer, and at the same time takes into account the effect on the absorption line of the effect of the variation of the continuous opacity with wavelength.

We have seen that, through formula (5) the value of R_c depends on both κ_R and κ_ν and hence on the wavelength. R_c may be considered as the optical depth in the continuum which corresponds to *H*, the " height " of the reversing layer.

2. M. Hack and Otto Struve. *Stellar spectroscopy, normal stars*, pp. 122-123, Trieste, 1969. This was the last book written by Otto Struve (1897- 1963), just after Astronomy of the XXth century (with Velta Zebergs)

3. in *Handbuch der Physik*, vol. 50, pp. 274-398 . See p. 360.

If h is a geometrical depth in the reversing layer reckoned from the stellar surface, the element of optical depth $d\tau_v$ in the continuum may be written as $\kappa_v \, \rho \, dh$, and, according to formula (6),

$$dh = d\tau_v / \kappa_v \rho$$

As $\kappa_v \rho$ is independent of h in the S.-S. model, we see that in order that, according to formula (6), $H = R_c / \kappa_v \rho$, we have to write :

$$H = \int_0^{R_c} \frac{d\tau_v}{\kappa_v \rho} = \frac{R_c}{\kappa_v \rho}$$

Hence, H is the height of the reversing layer and R_c represents the latter's optical depth in the continuum.

REFERENCES

L.H. Aller, *Ap. J.*, 104 (1946), 347.

L.H. Aller, *The Atmospheres of the Sun and Stars*, 1st edition, New-York, The Ronald Press C°, 1953, 2nd edition 1963.

D. Barbier, *Handbuch der Physik*, 50 (1958), 274-398.

A.S. Eddington, *Monthly Notices R. Astron. Soc.*, 89 (1929), 620.

S.S. Huang and O. Struve, *Ap. J.*, 123 (1956), 231.

D.H. Menzel, *Ap. J.*, 84 (1936), 462.

E.A. Milne, *Monthly Notices R. Astron. Soc.*, 89 (1928), 3.

M. Minnaert, *Z. für Astrophysik*, 10 (1935), 40.

G. Münch, *Ap. J.*, 127 (1958), 644.

A. Pannekoek, *Publ. Astron. Inst. Amsterdam*, n° 4 (1935).

A. Pannekoek, *Monthly Notices R. Astron. Soc.*, 91 (1930), 139 and 519.

C. Payne, *Harvard Collection Observatory Monographs*, 1 (1925).

A. Schuster, *Ap. J.*, 21 (1905), 1.

W. Schütz, *Z. für Astrophysik*, 1 (1930), 300.

K. Schwarzschild, *Sitzungsberichte der königl. Preuss. Akad der Wissenschaften*, Berlin, 1914, p. 1183.

M.H. Wrubel, *Ap. J.*, 109 (1949), 66 and *Ap. J.*, 111 (1950), 157.

C.A. Young, *The Sun*, 1893.

APPENDIX

Derivation of " Minnaert's formula " for a Milne-Eddington atmosphere

In an plane-parallel atmosphere in LTE, the transfer equation of the monochromatic intensity $I_v(\mu, \tau_v)$ where μ is the cosine of the emergent angle ϑ, can be written :

$$\mu \frac{dI_v(\mu, \tau_v)}{d\tau_v} = I_v(\mu, \tau_v) - B_v(\tau_v)$$

We shall call τ_v the optical depth in a spectral line and τ_c the optical depth in the neighbouring continuum. Hence, at the surface of the star $(\tau_v = \tau_c = 0)$, the solution of the transfer equation in the line is :

$$I_v(0, \mu) = \int_0^\infty B_v(\tau_v) e^{-\tau_v/\mu} \frac{d\tau_v}{\mu}$$

while in the neighbouring continuum, the intensity $I_c(0, \mu)$ is :

$$I_c(0, \mu) = \int_0^\infty B_v(\tau_v) e^{-\tau_c/\mu} \frac{d\tau_c}{\mu}$$

Let us suppose that we adopt a Milne-Eddington model with a surface temperature T_0 and let us represent $B_v(\tau_c)$ by a Mc Laurin series at the vicinity of the star's surface :

$$B_v(\tau_c) = B_v(T_0)\left[1 + \frac{1}{B_v(T_0)}\left(\frac{\partial B_v}{\partial T}\right)_{T_0} \frac{dT}{d\tau_R} \frac{d\tau_R}{d\tau_c} \tau_c\right]$$

where τ_R corresponds to the mean Rosseland absorption coefficient

Since

$$B_v(T) = \frac{2hv^3}{c^2}(e^{hv/kT} - 1)^{-1}$$

we shall write :

$$\frac{\partial B_v}{\partial T} = \frac{2hv^3}{c^2} \frac{hv}{kT_0^2} e^{hv/kT_0}(e^{hv/kT_0} - 1)^{-2}$$

In the Milne-Eddington model :

$$T_0^4 = T_{eff}^4/2$$

Then, according to the temperature law , where τ_R is a mean Rosseland optical depth :

$$T^4(\tau_R) = T_0^4(1 + 3/2\,\tau_R)$$

we may write by derivation :

$$\left(\frac{dT}{d\tau_R}\right)_0 = \frac{3}{8}T_0$$

On the other hand, in the M.-E. model, the ratios :

$$\frac{d\tau_R}{d\tau_c} = \frac{\kappa_R}{\kappa_c}$$

where the κ's are the absorption coefficients by unit mass, are independent of the depth in the atmosphere.

We can thus write :

$$B_\nu(\tau_c) = B_\nu(T_0)\left[1 + \frac{2h\nu^3}{c^2}\frac{e^{h\nu/kT_0}}{(e^{h\nu/kT_0}-1)^2}\frac{h\nu}{kT_0^2}\frac{3}{8}T_0\frac{\kappa_R}{\kappa_c}\tau_c\frac{1}{B_\nu(T_0)}\right]$$

With χ_0 given by expression (6) of the text, we can write :

$$B_\nu(\tau_c) = B_\nu(T_0)\left[1 + \chi_0\frac{\kappa_R}{\kappa_c}\tau_c\right]$$

Hence, the emergent intensity in the continuum will be written :

$$I_c(0,\mu) = B_\nu(T_0)\int_0^\infty\left(1 + \chi_0\frac{\kappa_R}{\kappa_c}\tau_c\right)e^{-\tau_c/\mu}\frac{d\tau_c}{\mu}$$

and, letting $\tau_c/\mu = z$,

$$I_c(0,\mu) = B_\nu(T_0)\int_0^\infty\left(1 + \chi_0\frac{\kappa_R}{\kappa_c}z\right)e^{-z}\ dz$$

As

$$\int_0^\infty(1 + \gamma z)e^{-z}dz = 1 + \gamma$$

the emergent intensity in the continuum is simply :

$$I_c(0,\mu) = B_\nu(T_0)[1 + \mu\chi_0(\kappa_R/\kappa_c)]$$

Integrating over the stellar disk, we find the flux in the continuum at the surface $F_c(0)$:

$$F_c(0) = 2\int_0^1 I_c(0,\mu)\mu d\mu$$

Or,

$$F_c(0) = B_\nu(T_0)[1 + (2/3)\chi_0(\kappa_R/\kappa_c)].$$

In a spectral line, the continuum opacity κ_c is replaced by the total opacity $\kappa_c + l_\nu$; hence, the emergent flux in the line is given by :

$$F_\nu(0) = B_\nu(T_0)[1 + (2/3)\chi_0(\kappa_R/(\kappa_c + l_\nu)]$$

Let us now write with theses expressions for the flux the observed line depth R_ν :

$$R_\nu = 1 - \frac{1 + \frac{2}{3}\chi_0\frac{\kappa_R}{\kappa_c + l_\nu}}{1 + \frac{2}{3}\chi_0\frac{\kappa_R}{\kappa_c}}$$

Simplifying the writing with $(2/3)\chi_0 = y_0$, we find, after some algebra,

$$R_v = y_0 \kappa_R \frac{l_v}{(\kappa_c + y_0 \kappa_R)(\kappa_c + l_v)}$$

At the center of a very strong line, l_v compared to κ_c and R_v will attain its maximum value we have called R_c in the main text. Hence, we find for R_c the following expression :

$$R_c = \frac{y_0 \kappa_R}{\kappa_c + y_0 \kappa_R}$$

or

$$R_v = R_c \frac{l_v}{\kappa_c + l_v}$$

Finally, this expression is equivalent to :

$$\frac{1}{R_v} = \frac{1}{R_c} + \frac{1}{R_c l_v}$$

Let us now show that this expression, in the case of the S.-S. model is equivalent to Minnaert's formula.

Recalling that in the Schuster-Schwarzschild model, the atmosphere is homogeneous, and letting N be the number of absorbers per unit volume, we find that l_v the line absorption coefficient per unit mass may be written :

$$l_v = \alpha_v \cdot (N/\rho)$$

ρ being the density of stellar material and α_v the atomic absorption coefficient.

Then, we can rewrite the expression for R_v as :

$$\frac{1}{R_v} = \frac{1}{R_c} + \frac{\rho \kappa_c}{R_c \alpha_v N}$$

which is our formula (7).

Cosmology and Technology

Gustaaf C. Cornelis

Big bang cosmology

Big bang cosmology originated out of mainly theoretical research (Einstein, Friedman, Lemaître, de Sitter, Robertson, and others). One can prove that at that time, during the twenties and thirties of this century, only Edwin Hubble's astronomical program eventually offered the empirical input the theoreticians could not neglect. In 1914, Vesto Slipher had shown that most of the so-called spiral nebulae (later identified as galaxies) regressed from earth, based on measurements of the redshift. However, Einstein and his followers (the theoretical cosmologists) did not know of this work when they took off with their theoretical research. So only after Hubble had found out that there is a correlation between the distance of galaxies and their redshift (in other words, the greater the distance, the greater the velocity), this empirical " evidence " for a dynamical universe was generally accepted.

There was sufficient empirical material to confirm the possibility that the universe could have come out from a rather small region. Evidently, it was Lemaître's work that had provided the necessary frame wherein afterwards the empirical data could fit.

Only a minority of astronomers and other scientists showed any interest in cosmology until the late fifties, because, according to the majority, it was too speculative. George Gamow was one of the exceptions. This Russian nuclear physicist, known for his generally accepted explanation for radioactive decay, interrelated the cause for the general expansion with the formation of the chemical elements[1].

---

1. G.A. Gamow, " Expanding Universe and the Origin of the Elements ", *Physical Review*, 70 (1946), 572-573 ; G.A. Gamow, " The Evolution of the Universe", *Nature*, 162 (1948), 680-682 ; G.A. Gamow, " The Origin of Elements and Separation of Galaxies ", *Physical Review*, 74 (1948), 505-506.

It was known by then that stars produce helium by fusion out of hydrogen atoms, only a fraction though of the helium measured in the universe — this is called the helium-abundancy problem. Most of the helium had to come into being during the pre-stellar period (before there were stars in the universe) when the unstable neutrons got coupled. Some neutrons decayed into protons, other formed helium with couples of protons.

This scenario was acceptable, given a temperature of one billion degrees[2]. Gamow concluded that the early universe must have been very hot and very dense. Sir Arthur Eddington had acknowledged twenty years earlier that helium could only be generated under extreme conditions : " The helium [...] must have been put together at some time and some place. We will not argue with the critic who urges that stars are not hot enough for this process ; we tell [him to] go and find a *hotter place* "[3].

Obviously, he did not come to the conclusion that helium therefore had to be formed in the universe before stars came into being. He did not put one and one together.

Gamow did[4]. Tolman had shown in 1931 that the dynamical models of the universe implied very high temperatures during the age of radiation dominance. Gamow pictured the universe as a thermonuclear fireball, a dense gaseous mixture of protons, neutrons and high-energetic photons he called " ylem " after the Greek word " ὕλη ", meaning " matter ". Under the extreme conditions, the heavier elements got formed in a chain of nuclear reactions.

Gamow's first proposal did not work. He found out that the expansion would not leave enough time to get all the helium and heavier elements we see around us nowadays to form. Gamow knew helium had to be formed in advance, but there seemed not enough time to do it, let alone to make the other and heavier elements.

Ralph Alpher and Robert Herman came to the rescue. These two students of Gamow found out that the main problem for the reaction sequence Gamow had proposed had everything to do with the " transition atoms " with an atomic mass between 5 and 8. These elements proved to be very unstable, hence, they couldn't melt together to form heavier and more stable atoms. Not like Gamow had described the process anyhow. When a proton and neutron were " glued " into a helium core, the resulting element would immediately fall apart. During the fifties, Alpher, Herman and Follin succeeded in developing a complete and

2. G.A. Gamow, " Expanding Universe and the Origin of the Elements ", *Physical Review*, 70 (1946), 572-573.

3. A.S. Eddington, *Internal Constitutions of the Stars*, Cambridge, Cambridge University Press, 1926.

4. This does not imply that Gamow did his initial research proceeding from a genuine cosmological question. In other words, his work belonged to another " research-program ".

adequate chain of reaction that took care of the helium abundancy problem and the formation of the heavier elements.

The interdisciplinarity between particle physics and cosmology proved to be very fruitful in the forties and fifties. Through the work of Gamow, cosmology could benefit immediately of the findings of particle physicists, heavily depending on the technological progress. This was crucial, though. Without the successes in that discipline, cosmology would probably not have been able to find that fast a solution for the helium abundancy problem. Of course, cosmologists would have started their own particle physics research programs, like they did later on, concerning the so-called " missing matter " and the Higgs boson, developments on which we will concentrate in the next part of this paper.

MISSING MATTER[5]

The overall geometry of the universe has to be either flat or curved. If it is curved, then the universe is open or closed.

There are reasons to " believe " that the universe is flat, which means that there's enough mass available in the universe to flatten it, to counteract the expansion. It is important to see that many scientists believe in the flatness of the universe ; it is not a question of knowing, but a question of believing. The reasons they put forward in favor of a flat universe are very speculative and no single proposal is conclusive. However, these do not concern us here.

In order to conclude that the universe is flat indeed, 90 % of the necessary amount of matter has to be found. In other words, the existence of ten times the amount of mass yet accounted for in the universe has to be shown. Only 1 % is actually seen : the stars and ionized gasses. Another 9 % can be indirectly proven to exist. Studies of galaxies show that there have to be great amounts of mass distributed all over the galactic disk to account for the velocity distribution. 90 % seems to be unobservable. Hence, the problem of " dark matter ", or better still : of " imperceptible matter " or " missing matter ".

This kind of matter, if it exists, has to be new and/or unseen. In the latter case, it concerns known sorts of particles which for one reason or another cannot be detected easily, therefore, cannot be readily perceived. Some cosmologists thought that heavy neutrino's were the particles that everyone was looking for. They were swift and although they were supposed eventually to be massless, one could suggest a fairly high mass for them.

However, evidence showed that this hypothesis was not working. Another possibility pertained to the axion. The axion was conceived in 1977 when cer-

5. French astronomers have concluded recently that the galaxy contains no dark matter at all. If it is there, it is certainly not in the galactic disks (as was suggested before). *Science*, 278 (14 November 1997), 1230.

tain problems in particle physics were piling up. These problems could be solved by the supposition of the axion. In a cosmological context, axions are plentiful ; there are even more axions then neutrinos. Hence, it would suffice for the missing matter problem that axions should have a mass of 100 000th part of one electronvolt. In comparison, heavy neutrinos should have a mass of 17 000 electronvolt. Axions are very interesting from a cosmological point of view. As particles, they are very stable, so they would have survived the big bang ; without this stability, these particles would have been already disappeared. From a theoretical point of view, the hypothesis that axions would make up the missing matter is quite plausible, because they were conceived in a complete different context. However, until now, axions are merely theoretical, since no empirical evidence has been found to prove their existence.

Regarding neutrinos and axions, the interdisciplinarity is twofold. Particle physics implies the existence of certain particles which can be implemented in cosmological theories to solve acute problems. On the other hand, some cosmologists search for certain particles, whose existence can only be proven if particle physicists have the necessary instruments at their disposal. It is obvious that the progress in technology in the field of particle physics has its influence on cosmological thinking. Without the technological progress, the particle physicists would not have conceived certain particles, which could help out the cosmologists. And without the technological progress, the existence of these particles, implied by cosmological theories and physical theories alike, cannot be proven. It is interesting to see that cosmological theories determine some features of the particles found through experiments done by particle physicists.

Axions and neutrinos are particles known long before the missing matter problem emerged. Yet, there are particles especially conceived to do something about the paradox. These were called fotinos, neutralinos or gravitinos. They should have masses between 10 and 10 000 proton-masses and a very low annihilation frequency, needed to survive the big bang. These two features would make them easy to find, given that they are abundant. This is obviously not the case. Henceforth, to prove their existence, they have to be recreated in a so-called accelerator, cyclotron or collider. Axions are quite heavy, so in order to recreate them under controlled circumstances, the energy needed to put into the system is equal to the energy of the particle in question (energy is proportionate to mass). No harm done, one should think, because new accelerators can always be planned and built. Scientists succeeded in this during the previous decades, so why should they not succeed in constructing a collider which produces the particles we are looking for ?

Indeed, cosmology initiated a program that had to be executed by particle-physicists. The characteristics of the subatomic particles to be found are such that new and special technologies were to be developed. The particle physicists did their job : the so-called *Superconducting Super Collider* was designed, even a site in Texas was assigned to it.

The Higgs boson

This instrument, the *Superconducting Super Collider* — now we shift our attention to the context of the Higgs boson — would also prove to be very useful regarding the answer to another cosmological question. This concerns the inflation model.

During the eighties, American and Russian scientists alike independently devised the so-called inflation models. To be brief, these pertain to a rather short and early phase in the development of the universe. Inflation theory implies an exponential growth of the universe, only a fraction of a second (10^{-35} s) after the big bang and which took about that period of time to take place. The inflation model solves a lot of cosmological paradoxes — flatness-, horizon-, smoothness-, monopole-problem — although only one — the monopole-problem — led to the development of the theory.

The monopole-problem belongs to the realm of particle physics, in contrast to the other paradoxes (which concern spacetime and the distribution of matter). The monopole-problem came about in the mid-eighties, as a result of the great unification theories. These theories concern the problem how the different forces in nature (strong and weak nuclear force, gravitation, electromagnetic force) can be considered as one. The unification theories group forces two by two. For example, the electromagnetic force combines magnetic and electric forces, as proved by Maxwell in 1865 (Faraday discovered the symmetry between the two forces). The main principle of the great unification theories is the thought that any force is merely a manifestation of one unified force. There is an underlying symmetry in the models that describe the forces. It is then believed that this unified force broke down into the four forces we know of, only a fraction of a second after the big bang. This process is called " symmetry breaking ". The theories are fruitful and compatible with the standard big bang theory, but they imply the existence of magnetic monopoles, particles not yet seen.

Symmetry breaking presupposes the existence of the Higgs boson, a very heavy particle, whose energy equals the energy needed to uphold the local symmetry and thus the unification of the forces. The symmetry breaking process is supposed to be spontaneous and local : so it does not have to occur everywhere at the same time and is not completely predictable. It is possible to keep up the symmetry longer than implied by the circumstances. In this case, one speaks of supercooling. It means that an enormous amount of energy is accumulated, which only gradually is transferred to the surroundings of the place where the symmetry-breaking takes place. A consequence of this process is that the universe undergoes a phase of exponential inflation during a very short period : the universe doubles its size every 10^{-34} s. It is clear that the findings in the field of particle physics have far-reaching consequences for cosmology.

Although the inflation model is quite successful theoretically speaking, it lacks any empirical support. The existence of the Higgs boson could change this. It is here that the *Superconducting Super Collider* would have served well.

However, in November 1993 the involved governments (British and American) started to ask clear legitimization for the funding (especially William Waldegrave), because the costs to build this accelerator are indecent. As a answer to this, many scientists held a plea (Steven Weinberg wrote his *Dreams of a Final Theory*[6] in this context), but none was conclusive. The tax-payer's money could and would not be spent on a (to the layman seeming) wild goose chase. The U.S. House of Representatives decided to halt the project after 14 miles of tunnelling were completed and already two billion dollars spent.

Although this particular story came to an end, in Japan new plans are developed to build a comparable collider. CERN too has new ideas for the future. One wonders whether the physicists will again draw attention to the cosmological significance of their research to make their plans acceptable.

CONCLUSION

In this paper, cosmology and particle physics are treated as " different " disciplines. Indeed, one can obviously speak of interdisciplinarity between these two sciences. There is no doubt that — methodologically speaking — cosmology and particle physics are quite different. For example, cosmology is speculative to a great extent. That is a reason for other physicists to be reluctant towards the cosmologist's " findings ". Another reason is the fact that cosmologists do not make a big deal out of numeral accuracy (according to a majority of physicists)[7]. For the particle physicists this is " inadmissible " for a legitimate science. Hence, there is an objective distinction between the two disciplines as well as a subjective difference.

The history of modern cosmology shows subsequently that the development of standard big bang cosmology heavily depended on the development of technology in the field of particle physics. In later years, the interdisciplinary influence became bilateral.

On one hand, cosmology implies new technologies (in particle physics). On the other, new technological progress expedites, or at least has a positive influence on cosmological research. The sociological importance of this peculiar interdisciplinarity is also clear : particle physicists (although many of them do not like cosmology at all) do take cosmology to legitimize their research to laymen. Of course, it is obvious that the public is intrigued by the big questions, those pertaining directly to the origin of the universe. The most important conclusion, though, is the fact that we are dealing here with two

---

6. S. Weinberg, *Dreams of a Final Theory*, London, Hutchinson Radius, 1993.

7. S. Perkowitz called my attention to this fact.

methodologically different disciplines which strongly influence each other on technological progress and by way of technology.

REFERENCES

A.S. Eddington, *Internal Constitutions of the Stars*, Cambridge, University Press, 1926.

G.A. Gamow, " Expanding Universe and the Origin of the Elements ", *Physical Review*, 70 (1946), 572-573.

G.A. Gamow, " The Evolution of the Universe ", *Nature*, 162 (1948a), 680-682.

G.A. Gamow, " The Origin of Elements and Separation of Galaxies ", *Physical Review*, 74 (1948b), 505-506.

S. Weinberg, *Dreams of the Final Theory*, London, Hutchinson Radius, 1993.

CONTRIBUTORS

Peter BARKER
The University of Oklahoma
Norman, OK (USA)

Javier BERGASA LIBERAL
Pamplona (Spain)

Luís Miguel CAROLINO
Universidade de Evora
Evora (Portugal)

Raz Dov CHEN-MORRIS
Tel Aviv University
Tel Aviv (Israel)

Gustaaf C. CORNELIS
Vrije Universiteit Brussel
Brussel (Belgium)

Suzanne DÉBARBAT
Observatoire de Paris
Paris (France)

Simone DUMONT
Observatoire de Paris
Paris (France)

Mark DE MEY
University of Gent
Gent (Belgium)

Erwin DE NIL
University of Gent
Gent (Belgium)

Martine GROULT
Ecole Normale Supérieure
de Fontenay-Saint-Cloud
Fontenay-aux-Roses (France)

Philippe HAMOU
Université Paris X
Nanterre (France)

Léo HOUZIAUX
Académie Royale de Belgique
Bruxelles (Belgique)

Elaheh KHEIRANDISH
Dibner Institute for the History
of Science and Technology
Cambridge, MA (USA)

Nicholas KOLLERSTROM
London (United Kingdom)

Alexander V. KOZENKO
Institute of the Earth Physics
Moscow (Russia)

Antoni MALET
Universitat Pompeu Fabra
Barcelona (Spain)

Hideto NAKAJIMA
University of Tokyo
Tokyo (Japan)

Dennis RAWLINS
Baltimore, MD (USA)

William R. SHEA
Université Louis Pasteur
Strasbourg (France)

Gérard SIMON
Université Charles de Gaulle, Lille 3
Villeneuve d'Ascq (France)

A. Mark SMITH
University of Missouri, Columbia
USA

Sabetai UNGURU
Tel Aviv University
Tel Aviv (Israel)

Ken'ichi TAKAHASHI
Kyushu University
Japan